Critical Path Method Tutor for Construction Planning and Scheduling

About the Author

E. William (Bill) East is an internationally known construction management researcher whose career spans the areas of planning and scheduling, quality management, and building information modeling. While with the U.S. Army Corps of Engineers, East created the Corps' Standard Data Exchange Format (SDEF). He also developed parametric planning and automated schedule analysis tools, and provided related consulting services on Corps' and Army projects. East's Design Review and Checking System (DrChecks^{SM}) streamlined design quality management for public project stakeholders at the federal, state, and local levels. Most recently, East invented the Construction-Operations Building information exchange (COBie)—an international standard that squeezes profits from an otherwise wasteful construction handover process. East has degrees in Civil Engineering from Virginia Tech (B.S.) and the University of Illinois (M.S. and Ph.D.). East is a registered Professional Engineer in the state of Virginia and is a Fellow of American Society of Civil Engineers (ASCE).

Critical Path Method Tutor for Construction Planning and Scheduling

E. William East

McGraw Hill Education

New York Chicago San Francisco
Athens London Madrid
Mexico City Milan New Delhi
Singapore Sydney Toronto

Cataloging-in-Publication Data is on file with the Library of Congress.

McGraw-Hill Education books are available at special quantity discounts to use as premiums and sales promotions, or for use in corporate training programs. To contact a representative please visit the Contact Us page at www.mhprofessional.com.

Critical Path Method Tutor for Construction Planning and Scheduling

1 2 3 4 5 6 7 8 9 0 QVS/QVS 1 2 0 9 8 7 6 5 4

ISBN 978-0-07-184923-4
MHID 0-07-184923-8

The pages within this book were printed on acid-free paper.

Sponsoring Editor
Robert Argentieri

Editorial Supervisor
Donna M. Martone

Acquisitions Coordinator
Amy Stonebraker

Project Manager
Ruchika Abrol, MPS Limited

Copy Editor
Surendra Shivam, MPS Limited

Proofreader
Garima Sharma, MPS Limited

Indexer
Diana Witt

Production Supervisor
Pamela A. Pelton

Composition
MPS Limited

Art Director, Cover
Jeff Weeks

For John Melin

Contents

Preface

When I started my career as a U.S. Army Corps of Engineers project engineer, I found that neither formal education nor available training prepared me for the day-to-day business of using Critical Path Method (CPM) schedules to evaluate progress payments, determine the impacts of contract modifications, or conduct claims analysis. After moving from the field to the Construction Engineering Research Laboratory, I had the opportunity to use and grow my experience through consulting and schedule analysis services for projects both large and small. One of these projects was the Standard Data Exchange Format (SDEF) for project scheduling. Contractors have been submitting their monthly progress schedules to the Corps of Engineers using SDEF for the last three decades.

When asked to join the faculty at the University of Illinois at Urbana-Champaign, I jumped at the chance to share my hands-on experience. I updated the project planning part of an introductory course in construction management to simply explain CPM and provide practical explanations of why and how specific scheduling practices were helpful. With the later advent of the Internet, notes from that class became the popular CPMTutor.com website. Over the years, practitioners, students, and educators visiting CPMTutor.com asked me to develop my materials into a book. You are looking at the result of those requests.

The effective and consistent use of the techniques introduced in this book will help the reader to:

- Reduce progress payment disputes and speed progress payment
- Plan resources for maximum productivity
- Predict owner-caused and weather delays
- Develop "get well" plans for projects that fall behind
- Correctly update progress, including out-of-sequence activities
- Meet contracts requiring CPM schedules
- Evaluate change and claim impacts; develop admissible documentation

While some may be more inclined toward planning than others, planning is an essential human activity that anyone can learn. I hope that the approach taken in this book, to explain both the "why" and the "how" of CPM scheduling, will be helpful in your quest to bring every project on time and under budget.

Thanks to Dave Quigley for helping me find a home for this book at McGraw-Hill Education and for his helpful manuscript reviews. Also thanks to Dana Finney, Kathryn East, and Ruchika Abrol for their help with copyediting. Finally, thanks to those of you who have contacted me through the CPMTutor .com asking how to get this book!

E. William East, Ph.D., P.E., F.ASCE
Prairie Sky Consulting
November 2014

A Note on Typography: *When referring to activities on a construction schedule, this book will always surround the name of the activity with quotation marks. For example the activity named "Task 1" is always shown in quotes as the activity named "Activity A" or "Backfill."*

Many terms in the Critical Path Method (CPM) have acronyms that are widely used in this book. In cases where these terms do not have acronyms these terms will remain capitalized, as if they were proper nouns. This is done to emphasize and reinforce the definitions of these specific terms and phrases.

CHAPTER 1

Why Scheduling?

The need to formally schedule projects often starts without a proper introduction. For example, a small contractor gets their first job with a new, large owner. That owner's contract requires a Critical Path Method (CPM) schedule. In other cases, a new engineer at a general contractor's office may be told to create a computer schedule without knowing how to go about doing it. Due to time constraints, students may only have the opportunity to learn the mechanics of scheduling. In each of these cases, the person creating and using the schedule can easily miss the reason for making a schedule in the first place.

This chapter explains what a construction schedule is and is not, and describes the benefits that can be achieved through the use of an effective construction schedule. Understanding these basics before applying the mechanics of scheduling will help the reader to understand the purpose, value, and limitations, of the techniques presented in this book.

What Is a Construction Schedule?

A construction schedule is a model of the steps required to complete a construction project. It is very unlikely that two construction schedules will be exactly the same because no two construction projects are exactly the same.

A construction schedule is not an exact model of a construction project. Like any other model, a schedule approximates some aspects of a construction project and completely omits others. If the scheduler understands the trade-offs made when creating the schedule, then the schedule can be a very useful tool indeed.

Effective project schedules include: (1) the sequence of activities needed to complete the work, (2) the duration of those activities, (3) the flow of crews through the project who perform those activities, and (4) the value generated by the completion of each activity toward the overall completion of the project.

Rather than simply using a schedule to meet a contractual requirement, effective scheduling helps the project team anticipate and resolve potential problems before they occur. The effective use of schedules also helps the team quickly evaluate and resolve problems that ultimately arise on large or complex construction projects. The paragraphs below discuss the specific questions that can be answered if the project team adopts and uses scheduling best practices.

Will the Project Finish on Time?

The most common reason that owners include schedules in construction contracts is that owners assume that the initial schedule and monthly updates will present an accurate account of work completed and a prediction of work remaining. Schedules developed for the sole purpose of meeting owners contract requirements are often created by off-site personnel. If sufficient effort is made to discover on-site conditions and subcontractor constraints, then these third-party schedules may result in good initial schedules. Initial schedules help to determine if a construction sequence will be able to complete the project with the allotted duration.

After the submission of the initial schedule, the schedule is required to be updated monthly to identify efforts toward completion. Schedule updates that take account of local site and job conditions can be good predictors of the ability of an ongoing construction project to complete the project within the remaining time.

Is the Project on Schedule?

If a contract requires a construction schedule, it is likely that the schedule will also be used as a basis for periodic progress payments. Often the use of a schedule in this case is simplified to be a "schedule of values" without regard to the sequence and physical progress of each individual activity. If, however, the schedule is used correctly, the value of in-progress and completed activities can help to identify differences between the project plan and actual performance. Correctly defined schedules track both the time and cost (or earnings) completion of each activity. This means that schedule activity updates must reflect those tasks that are actually taking place on the jobsite, not the initial "schedule of values." Correctly defined in-progress activities report two values each month: "Remaining Duration" and "Percent Complete."

Separating time from cost allows activities that have large up-front equipment costs to correctly accrue earnings. The day after delivery of a generator the "Install Generator" activity may have, for example, a Remaining Duration equal to the Original Duration and a Percent Complete of 75%.

Another way that the correct evaluation of in-progress activities can occur is to identify work that is falling behind schedule. Consider an activity, such as wall painting, that typically uses the same amount of resources each day. For such activities we could expect that if 75% of the value of the work has been earned, then the time required to complete the activity would be about 25% of its Original Duration. If, however, the owner paid for 75% of the activity's value, but it was estimated that the activity had 75% of its duration left to complete, then the activity is in trouble.

What Crews Should Be on Site Today?

On complex projects, one of the most difficult coordination efforts is ensuring that subcontracting crews arrive in a sequence that allows trades to follow one another. Where many trades are present, one trade out of sequence can result in rework. The coordination required is not a technology problem, nor a problem that requires new programs and management; it is simply a planning and communication problem. The communication can be made more clear through the use of schedules that model crews' flow through a project.

If schedules specifically identify subcontractor activities, then the timing for the arrival, scope of work, and departure of subcontractor crews can be clearly described to all stakeholders. Using the schedule as a communication vehicle can help decrease miscommunications. If there are delays or scheduling conflicts, then the impacts of these delays on following contractor and/or subcontractor activities may be clearly identified.

What Happens If?

If the initial schedule accurately reflects the way that the prime and subcontractors plan to do the work, and the schedule is updated as the project goes along, then the schedule can predict what may happen if changes are introduced to the plan. Schedule changes will directly add or remove specific activities and may also impact other activities and the overall completion of the schedule. Having a mutual understanding and agreement on pending change requests during the course of the project is usually preferable to legal depositions and court appearances years after a project has completed.

What Happens Now?

The Google search on "Why bad things happen to good projects" turns up over 375 million hits! Anyone who has ever been on a construction site can write that book. A schedule that accurately reflects the initial project plan and is updated to reflect changes during the project allows the team to quickly identify and mitigate problems that arise. Following that, the impacts of these problems can be quickly addressed in appropriate contract changes before the project reaches the "claims phase."

What Happened?

If accurate initial and updated schedules are not properly maintained, it will fall to the claims team to answer the question of "what happened?" The team will use those schedules or schedules of value that were available, along with daily reports, jobsite photos, invoices, submittals, and payroll records to reconstruct an accurate picture of the construction schedule. Of course, doing this after everyone has left the jobsite and may not even be available to be interviewed is not the mark of an effective scheduling effort. Projects scheduled using the techniques described in this book should be able to avoid the grueling and nonproductive effort needed to resolve issues during the claims phase.

What Is a Control Cycle?

Imagine how long it would take to get a new car if the manufacturer simply delivered all the parts to each house and the workers had to travel to each location and assemble the car. This is exactly what happens on a construction site. Construction projects create a temporary factory on the jobsite. While premanufacturing of duct-runs and pipe-spools is being used more often, assemblies are still connected at the jobsite.

In the factory, it is possible to directly measure the productivity, quality, and cost of each step of the manufacturing process. On the construction site, the project team relies on the quality control, quality assurance, standard and repetitive detailing, and personal experience to ensure that the job is done to the level of quality required by the construction contract. Over time, construction management tools have been developed to assist in these tasks. These tools are applied as part of "control cycles."

Control cycles allow managers to define objectives for a specific aspect of the project, measure the performance of the work as it proceeds, evaluate the work against the predefined objectives, and decide if adjustments are needed to resolve any difference between plan and actual. Effective project managers agree that there are four different aspects of projects that require such control cycles: time, cost, quality, and scope.

The typical controls for each topic, time, cost, quality, and scope are outlined in tables below. In each table, the first column, Component, lists those aspects of a project typically managed under that control regime. The second column, Element, identifies the specific area of study needed to address the topic. The third column, Concerns, identifies the choices available to managers when faced with decisions regarding these elements. The forth column, Technique, lists techniques or methods used by project managers to help them make such decisions.

Table 1.1 identifies the Components, Elements, and Concerns related to time control. This table is the one directly related to the contents of this book. The two components of time control that must be addressed are Sequence and Duration. Of all the

Component	Element	Concern	Technique
Sequence	Sequential Parallel	Trade interaction Work area access Physical constraints Common practice	Fenced bar charts CPM PERT Line-of-Balance
Duration	Productivity	Technology choice Equipment choice Crew size	Seat-of-pants Bar charts

TABLE 1.1 Time Control

time-control techniques, CPM is the most widely applicable. Other time-control techniques are discussed in this book to highlight the importance and value of CPM.

Table 1.2 identifies cost control techniques often grouped together in “engineering economics” studies required of undergraduate civil engineering students. Students in business schools will also recognize many of these techniques as essential

Component	Element	Concern	Technique
Material	Direct costs	Price	Inventory control Just-in-time delivery Storage
	Indirect costs	Waste Shortages Handling cost	Example "panels" Fabrication detailing Laydown simulations Barcode/RFID tracking
Labor	Direct costs	Productivity Wage rates	Crew selection Crew size Subcontracting Overtime use
	Indirect costs	Mobilization Overheads	On-site storage Job conditions Breakeven analysis Learning curve
Equipment	Direct costs	Efficiency Productivity Operating cost Maintenance cost	Rent/lease/own Storage/transportation Tire management Service contracts
	Indirect costs	Depreciation	Rent/lease/own

TABLE 1.2 Cost Control

Component	Element	Concern	Technique
Materials	Life-cycle costs Maintenance costs	Incorrect application High operating costs High disposal cost	Codes and standards Specifications Submittal compliance
Installation	Tear-out Rework	Incorrect installation Trade conflicts Crew productivities	Training Crew sequencing Quality control

TABLE 1.3 Quality Control

elements of managerial accounting. The complete coverage of cost control is outside the scope of this book; however, since "time is money," those aspects of cost control that can be illuminated by construction scheduling will be introduced in this book.

Quality concerns, shown in Table 1.3, are the third type of control system put in place on most construction projects. The quality control system ensures that the work meets the quality requirements identified in the plans and specifications. Far from being an abstract idea, quality in construction is defined by the construction contract itself. Given the wide variety of materials, products, equipment, and crews on a jobsite, the question for a contractor is how to achieve the specified quality at the least cost.

The impact of time-control techniques on quality control is primarily a matter of trade sequencing. If, for example, both sides of an interior partition are completed prior to installing electrical conduit and plumbing, part of that newly built wall will have to be torn out and subsequently repaired. The scheduler needs to be sure that work is accomplished with as little tear-out and rework as possible. To reduce tear-out and rework, activities will generally be scheduled using an "inside-out" sequence. In other words, the work to be covered up comes before the work that covers. A practical example in a mid-rise office building is that the heating, ventilating, and air-conditioning (HVAC) ductwork above the ceiling is installed before the ceiling grid is installed. Once the grid is installed, the diffusers and flexible connections to the diffusers can be installed. To install the ceiling grid and tiles before the HVAC system would result in significant tear-out and rework.

Component	Element	Concern	Technique
Scope	Features of work Bid items	Scope creep Mission change	Design review Cost growth management Change order oversight

TABLE 1.4 Scope Control

Another type of control requirement often ignored by new schedulers is scope control. The elements, concerns, and techniques pertaining to scope control are found in Table 1.4. Scope control refers to the content and timing of the work to be accomplished in the contract. For example, if the owner wants more work to be included on the project, a scope change is needed. If the owner changes the quality or other general provisions in the contract, then a scope change has also occurred.

Time-control techniques are essential to analysis of changes to a project's scope. To use time-control techniques effectively, a good scheduler will know the current status of the project, be able to predict the progress of the project if the change did not occur, and be able to estimate the changes to the project's progress given specific changes to the project.

Controlling time, cost, quality, and scope require extensive knowledge, experience, and effort. Such effort does not come without cost that must be added, as overheads, to a construction project's budget. Some project owners, contractors, and subcontractors do not develop schedules precisely because they have to budget to cover the cost of scheduling.

Decisions about which control systems to put in place, and which not to put in place, can be resolved by answering two questions. The first is, "Does the contract require it?" Most contracts require specific control systems, and may even mandate the use of specific software or software standard exchanges and reports. Since the owner is paying for the contract, including general provisions, then the owner has determined it is in their own interest to require specific control procedures.

When a specific control system is not included in the contract, the prudent contractor must also ask, "Will the investment in a given technique result in higher profit than would have been possible without it?" In most cases, the use of control systems on large or complex projects, as long as they are reasonably used, may result in higher profit resulting from faster payments from the owner and fewer cost over-runs. One of the purposes of this book is to help the project team to determine what constitutes a "reasonable use" of project planning techniques.

What Is the Critical Path Method?

The Critical Path Method (CPM) is a set of scheduling and modeling techniques that assist in planning, control, and prediction of a project's sequence, duration, and earnings. A CPM schedule breaks large, complex projects into activities whose duration and sequence can be measured. Given a list of activities, the arithmetic of CPM simply adds activities' durations, according to their sequence, to estimate the project's duration. The term "critical path" refers to the set of activities in a project whose total sequence is longer than any other set of activities. The length of the critical path is how long it is expected to take to complete the project.

How to Use CPM?

CPM could be seen as the simple application of the arithmetic rules needed to evaluate the duration and sequence of a set of activities. Such an assessment may be relevant from academic study; however, the blind use of CPM calculation without an understanding of how to effectively use it on a construction project will result in the project team ignoring or misusing the schedule.

To create an effective schedule, several steps are required. First, the scheduler should review the contract to identify all major milestones and features of work. Next, these requirements should be listed with their contractually required start dates, durations, and any required end dates. Due dates of administrative tasks that may delay construction should also be added to ensure potential impacts resulting from the construction manager or owner's representatives do not impede the project's completion.

The project is then decomposed into work areas. These work areas define the general areas where crews will flow through a project. Most projects will have a combination of work areas defined both horizontally on the site and vertically in the building or facility. The construction equipment and/or technology to be used should be considered since the choice of technology will necessitate the choice of the crew and productivity rates.

As crews move through each work area, the sequence of specific activities and durations can be defined. When the scheduler combines these activities using CPM, an initial project schedule is created. The schedule can then be adjusted based on contract or trade-based requirements. The size of crews may be adjusted to ensure that a consistent productivity can be maintained through the project. During this analysis, it can be determined whether it is possible to accelerate critical activities to decrease the overall cost of the project. The funding needed to finance the project may be defined. By the time the reader completes this book, they should be able to accomplish these tasks and ask specific questions of project partners and stakeholders if those answers are not clear.

This chapter began by posing a series of questions commonly asked of schedulers. The effective use of CPM can answer these initial questions and more. Some of the additional questions that can be answered by CPM are provided below:

- What is the impact of alternative construction equipment or crew sizing on the project?
- Which activities in a project control the overall completion of the project?
- Have all features of work been considered in the project plan?
- What financing is needed for the project?
- How can I best communicate the project plan to all team members?
- When will subcontractors need to arrive and depart from the site?
- When can the owner take beneficial occupancy?
- What is the expected cost of a change order?
- What is the impact of a set of unexpected contract changes?

CHAPTER 2

A History of Project Planning

The Egyptian pyramids were the first documented civil works project. Hieroglyphics describe the organization of work crews consisting of units from 200 men up to groups of 2000. According to Miroslav Verner in *The Pyramids: The Mystery, Culture, and Science of Egypt's Great Monuments*, archeological evidence clearly shows that the ancient Egyptians were the first construction managers.

Like the builders of today, pyramid builders maintained labor and material records. Archaeological research demonstrates that a significant effort was needed to create camptowns and to supply the meat, bread, beer, and onions needed to pay workers. People building these projects had some problems in common to those in our modern world. John Romer in *Ancient Lives: Daily Life in Egypt of the Pharaohs* documents strikes on several pyramid projects. In response to these strikes, a new class of scribes was created to manage project resources. Today, such scribes are called construction managers!

Modern planning can be attributed to the "scientific" process of factory production management. The pioneer of this approach was Frederick Taylor. Taylor's 1914 book, *The Principles of Scientific Management*, explains that his goal was to develop an approach to production that could eliminate strife between management and labor. This effort was claimed to improve both productivity and wages by detailed tracking of the time required for workers to perform each step of a manufacturing process.

Instead of uniting management and labor, the productivity controls resulting from Taylor's time tracking studies often increased labor tension as more work was required for little additional pay. With managers looking for ways to clearly communicate productivity goals, a series of graphical tools began to be used in factories. The most famous of these tools, from the point of view of the construction manager, was the Gantt Chart. In the 1922 book *The Gantt Chart: A Working Tool of Management*, the Gantt Chart is described as the "most notable contribution to the art of management made in this generation." As a factory production tool, a key to the use of the Gantt Chart was its ability to help coordinate the handoff of work among different types of machines and mechanics. To answer the problem of increased complexity, W. Edwards Deming would, in the middle of the 20th century, remark that most instances of poor worker productivity were caused by ineffective management control systems.

Today's project managers understand that they cannot blindly subscribe to the outputs of one, or even several computer systems as if the reports provided by those systems are truth. Rather than a rigid approach to planning, a nuanced approach is needed. Today we understand that scheduling systems, like any other system, only models some aspects of the construction process. The effective project manager takes to heart the famous quote by George Box, "All models are wrong; some models are useful." The quickest way for construction managers to understand the truth in this statement is to review this history of scheduling methods and to see where each works well, and where each begins to break down.

Bar Chart

The Gantt Chart, which is commonly called a Bar Chart, illustrates a list of activities and their durations. Table 2.1 provides an example list of activities and their duration. Such information is necessary, but insufficient to make a complete project plan. The third column of Table 2.1 completes the information needed—the sequence, or logic, of the schedule.

Activity	Duration (Workweeks)	Preceding Activity
Task 1	3	None
Task 2	5	Task 1
Task 3	2	Task 2
Task 4	3	Task 1
Task 5	2	Task 4

TABLE 2.1 Activity, Duration, and Sequence

Just as people may have preferences as to the name of the Gantt Chart or Bar Chart, there are synonymous ways to describe the logic of a schedule. There are even differences in the way that some prefer to list the sequence of activities in a project. Sometimes, an activity list will show preceding, or prior, activities. Sometimes, the logic of the schedule will be described by listing the succeeding, or following, activities. In the case of Table 2.1, the third column provides the list of preceding, or prior, activities. Ultimately, it is the job of the scheduler to communicate the results of whatever specific terminology is applied on a given project. An effective scheduler is one who is able to use the terminology that most clearly communicates with their team members.

The first column of Table 2.1 lists the activities needed to complete the example project. While there are no set rules about creating an activity list, Gantt preferred to measure in units of production that could be counted, such as "rivets driven," versus general measures of productivity such as percentage completion. Gantt also was keenly aware of what every experienced construction manager today knows—that as soon as a project starts, the activities needed to complete it may also need to change.

The second column lists activity duration. When discussing duration, the scheduler must clearly communicate assumptions about the construction calendar and units of duration. For example, the first activity in Table 2.1, "Task 1" has a 3-week duration. Note that there is no mention made of the length of that workweek. The total amount of effort for these 3 weeks will clearly be different if crews worked a 5-day or 7-day workweek.

As is the case with terminology surrounding the schedule's logic, the scheduler should clearly communicate the nature of the calendars on a schedule to ensure there are no ambiguities.

The third column lists the activities that must be completed before starting the given activity. This is the schedule logic or sequence. For example, prior to starting "Task 4," "Task 1" must be completed. In many cases, the sequence between construction activities is very clear. In other cases, the assumptions made in the schedule do not fully reflect the complexity of a construction site. Understanding when these assumptions do and do not apply will be critical to the effective use of any schedule.

The information in Table 2.1 can be used to create the Bar Chart shown in Fig. 2.1. The *x*-axis of the Bar Chart provides the project time line. The *y*-axis of the Bar Chart lists each activity. The contents of the chart are bars that illustrate the duration of each activity. The length of each bar is proportional to the duration of the activity. Activities on a Bar Chart are typically listed with the first activities at the top of the chart (the greatest *y*-axis value) and the last activities at the bottom of the chart (the least *y*-axis value). If the Bar Chart is organized in this way, the typical Bar Charts will show a pattern of bars that begin in the upper-left of the chart and proceed to bars displayed in the bottom-right.

The advantage of the Bar Chart is its simplicity. Figure 2.1 is quite easy to understand, simply by inspecting the graphic image. Consider, however, if the number of activities to be scheduled were over 50, over 200, or over 1000. In any large project, the corresponding Bar Chart would need to be spread across

Figure 2.1 Simple Bar Chart.

multiple pages in both the *x*- and *y*-directions. If Bar Charts are required for such large projects, then the scheduler should ensure that the activity ID number is printed above or inside the activity bar. On large charts, the *x*-axis should be printed at the bottom of every page, if only to help align pages as they are pasted together to create a single wall-sized Bar Chart.

The disadvantage of the Bar Chart is also its simplicity. The Bar Chart omits an entire category of information—schedule logic. In Fig. 2.1, for example, there is nothing on the Bar Chart itself that would keep someone from thinking that "Task 4" could be moved forward to start at week 1. Since we know that "Task 4" cannot begin until the conclusion of "Task 1," as shown in Table 2.1, the simple Bar Chart leaves significant room for improvement. Despite this problem, the Bar Chart remains a commonly used project planning tool.

Fenced-Bar Chart

John Melin and Barry Whiteaker addressed the question of how to fix Bar Charts in their 1981 paper, "Fencing a Bar Chart." The Fenced-Bar Chart includes vertical lines or "fences" to graphically add the schedule's logic directly on the Bar Chart. Figure 2.2 shows the Fenced-Bar Chart for the project described in Table 2.1. The first fence on the chart is found at the end of "Task 1." This first fence connects the completion of "Task 1" with the start of both "Task 2" and "Task 4." With the addition of this simple construct, the fence, the reason that "Task 4" cannot begin in week one is unambiguous. Any delay to "Task 1" will also delay the start of "Task 2" and "Task 4."

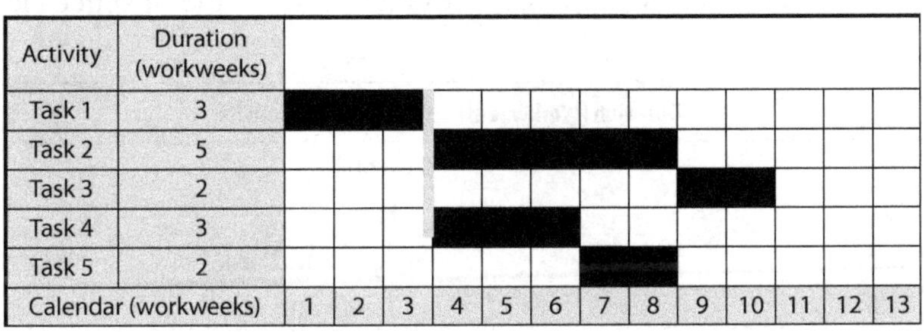

Figure 2.2 Fenced-Bar Chart.

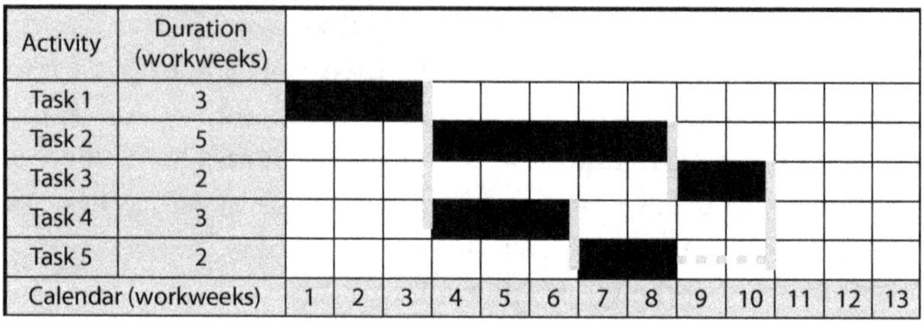

FIGURE 2.3 Parallel-logic Fenced-Bar Chart.

Adding the next fence in Fig. 2.2 starts at the end of "Task 2." This fence illustrates that the start of "Task 3" is constrained by the completion of "Task 2." A fence between "Task 4" and "Task 5" illustrates the proper sequence between these activities. To complete the Fenced-Bar Chart, the final activities in the schedule "Task 3" and "Task 5" are joined by creating an ending fence. The complete set of fences in Fig. 2.3 shows that "Task 3" completes 2 weeks later than "Task 5." This is illustrated by the dashed line that starts at the end of "Task 5" and extends to the ending fence, at the end of "Task 3."

The Fenced-Bar Chart in Fig. 2.3 can also illustrate the potential impact of changes in parallel paths of activities. "Task 2" and "Task 3"; and "Task 4" and "Task 5" could have different impacts on the completion of the project. Any delay to either activity "Task 2" or "Task 3" will push the project beyond the planned completion in week 10. On the other hand, a delay of 2 weeks or less to "Task 4" and "Task 5" will allow the project to be completed without delay.

Fenced-Bar Charts are extremely powerful and effective. Even projects with complex logic can be illustrated using Fenced-Bar Charts. Table 2.2 provides an example of schedule

Activity	Duration (Workweeks)	Preceding Activity
Task 1	3	None
Task 2	5	None
Task 3	2	Task 1 and Task 2
Task 4	3	Task 1 and Task 2

TABLE 2.2 Shared-Logic Schedule

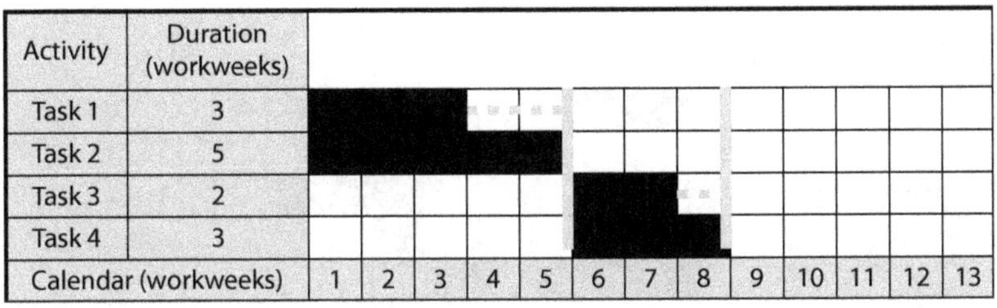

Figure 2.4 Shared-logic Fenced-Bar Chart.

where the sequence between the activities would be lost on a Bar Chart. As shown in Fig. 2.4, one single fence connects "Task 1" and "Task 2" to "Task 3" and "Task 4." Since the duration of "Task 1" is 2 weeks shorter than that of "Task 2," there is a dashed line that follows "Task 1" to connect it to the following activities. Similarly, "Task 3" is connected to the ending fence with a dashed line. Delays to activities without a dashed line can be shown to push back the entire project. Delays to activities with dashed lines will only delay the entire project if the duration of the delay is greater than the length of the dashed line.

Of course, not all schedules are as simple as those shown in Figs. 2.3 and 2.4. Typically, schedules will have activities where only part of the logic is shared. In Table 2.3, the logic of the schedule is more complicated because "Task 3," "Task 4," and "Task 5" do not share the same set of preceding activities. Figure 2.5 provides the Fenced-Bar Chart for Table 2.3.

The first thing to notice about Fig. 2.5 is that the activities are not listed in the same order as found in Table 2.3. In the

Activity	**Duration (Workweeks)**	**Preceding Activity**
Task 1	3	None
Task 2	5	None
Task 3	2	Task 1
Task 4	3	Task 1 and Task 2
Task 5	1	Task 2

Table 2.3 Overlapping Logic Schedule

FIGURE 2.5 Overlapping logic Fenced-Bar Chart.

figure, "Task 1" and "Task 3" are grouped together at the top of the chart. "Task 2" and "Task 5" are grouped together at the bottom of the chart. With this layout, the fence connecting "Task 1" and "Task 3," and "Task 2" and "Task 5" is easily interpreted.

In the middle of Fig. 2.5 is the activity, "Task 4," with overlapping logic. With a fence from both "Task 1" and "Task 2" preceding "Task 4" the proper sequence is maintained. If, for example, "Task 1" was delayed 3 weeks, then the only activities that would be affected would be "Task 3" and "Task 4." If, on the other hand, "Task 2" was delayed 1 week, then only "Task 4" and "Task 5" would be affected.

Fenced-Bar Charts are a very effective communication tool. These charts can explicitly illustrate both sequence and duration in a very intuitive way. Given the visual power of these charts, several advanced scheduling topics covered later in this book will employ them. For large construction projects, manual illustration proves to be as problematic for the Fenced-Bar Chart as the Bar Chart.

Line-of-Balance

Following the creation of Bar Charts, another graphic technique developed was called "Line-of-Balance" or LOB. LOB was developed by the Goodyear Aircraft Company and used extensively in planning the production of military equipment during World War II. According to G. Henrich and L. J. Koskela in their paper, "Evolution of Production Management Methods in Construction," the first construction industry application of LOB was ascribed to the United Kingdom's National Building Agency. In LOB, the x-axis, just like the x-axis of the Bar Chart, is a time line. Unlike the Bar Chart, the LOB's y-axis is the total number of units required to complete the project.

An activity in LOB is represented as a line. Figure 2.6 provides an example of an LOB diagram with one activity that is part of a seven story building project. The activity line's starting point is on the x-axis, on the day that the crew reports to work, and with a y-axis, value of zero units produced, since the activity is just starting. As the first activity in Fig. 2.6, "Task A" simply begins at the point (0, 0). The point for the end of the activity line has an x-axis coordinate of the last working day for that activity. The y-axis coordinate for the end of an activity is at the top of the chart, showing that the crew performing the work has fully completed all required units. In Fig. 2.6, the end of work for "Task A" will be at the point (14, 7). The slope of the line is the productivity of the crew performing the work. The crew performing "Task A" is planned to complete one floor every 2 weeks.

Adding the next activity to a LOB diagram begins by finding the end of the preceding activity at the first work area. In the

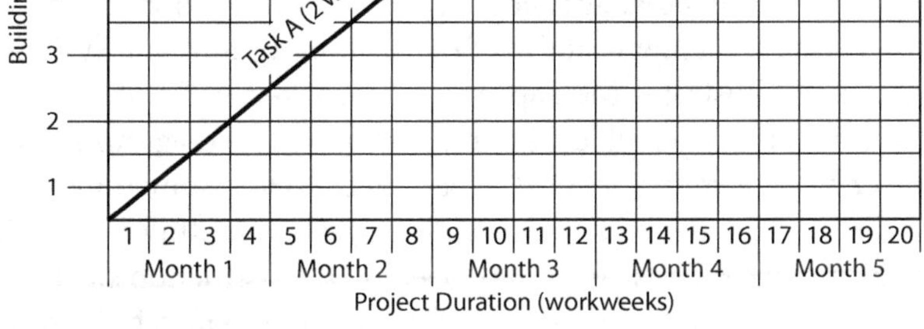

Figure 2.6 Line-of-Balance diagram.

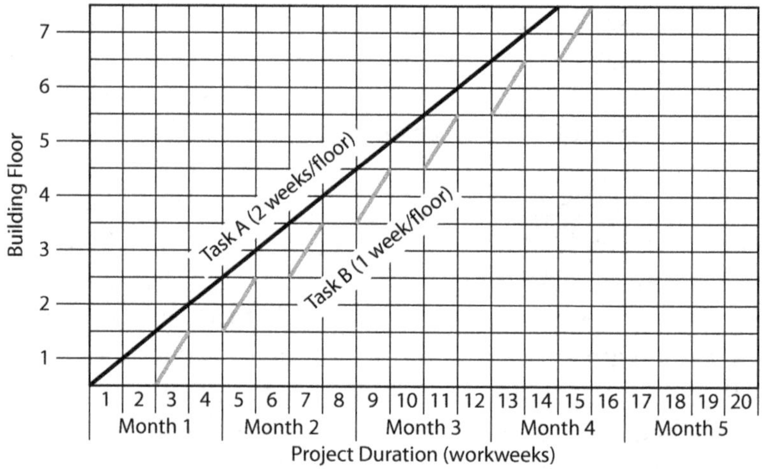

FIGURE 2.7 Line-of-Balance with two activities.

example project shown in Fig. 2.7, it is assumed that crews will follow each other starting at the first floor and moving through the project in-turn. Thus, the earliest date on which "Task B" could begin to work on the first floor is when "Task A" begins working on the second floor. Therefore, the starting point of "Task B" will be the point (2, 0).

To determine the end point of "Task B," the scheduler creates an activity line from the starting point for "Task B" and the slope, or production rate, for "Task B." The production rate for Task B is one week per floor. If the rate of production of "Task B" is less than that of "Task A," this procedure would be successful. If, however, "Task B" can complete its work faster than "Task A," then there are only two options. The first option is for "Task B" to immediately follow "Task A" on each floor, and then have "Task B's" crew wait for "Task A" to move to the next floor. Figure 2.7 illustrates the gap in work required in "Task B" if it starts as early as possible. The implication of this diagram is that the crew performing "Task B" would need to find something else to do, or delay their rate of production, while waiting for "Task A" to complete. Those who have been on a construction site have often seen crews slow their rate of production to match the productivity rate of preceding activities.

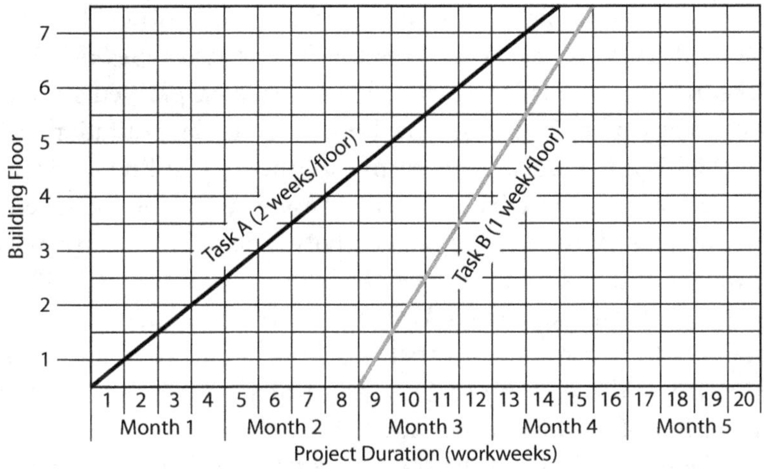

Figure 2.8 Line-of-Balance showing pacing activity.

Another approach to scheduling "Task B" would be to delay the start of "Task B" until its crews can begin the project and work through until the end of the job without delay. Figure 2.8 demonstrates a LOB schedule where the previous activity "Task A" sets the pace for the production of the following "Task B." If "Task B" waits to start until week 9, then the crew will be able to work at their proper productivity rate.

LOB has been applied to both vertical and horizontal construction and is still in use today. Any project that has multiple work areas through which crews "flow" or "chase" each other is a good candidate for the use of LOB. Even larger schedules that may have to be scheduled using computerized methods can be simply presented in an LOB diagram. LOB introduces the concept of the "crew chase" that will be covered later in this book.

Arrow Diagramming Method

To coordinate the work of larger, more complex projects and address the many nonproduction issues that may impact a construction project, J. E. Kelley and M. R. Walker introduced a new planning technique in their 1957 paper, "Critical-Path Planning and Scheduling." Originally called "activity-on-the-arrow" scheduling, and now commonly called the Arrow Diagram Method (ADM), this ADM was the first of two types of CPM scheduling methods. ADM directly addresses the problems of Bar Charts by explicitly representing both sequence and duration.

ADM introduced new terms to the scheduling lexicon. These terms are "activity arrow" and "node." In ADM, the scheduler creates one arrow for each activity. The tail of the arrow is the start of the activity. The head of the arrow is the end of the activity. When drawing an ADM schedule, each activity's arrow may be drawn in a length that is proportional to its duration. The use of these "time-scaled" ADM diagrams reinforces the metaphor of a project time line.

Nodes are placed at the start and end of each activity arrow. The starting node for a given activity is that activity's "i-node." The ending node for a given activity is identified as that activity's "j-node." This nomenclature is used because the mathematical model and associated computer implementation of ADM uses a summation symbol whose bounds are the variables "i" and "j."

Nodes are used to illustrate the schedule logic. To show that one activity follows another, the head of the preceding activity enters a node, and a following arrow starts at that same node. Since projects are defined by a specific start and end, there is a single starting "i-node" and a single ending "j-node" for each activity. Figure 2.9 shows the schedule from Table 2.1 in ADM format.

As can be seen from Fig. 2.9, ADM is intuitive because both sequence and duration are identified by the same activity arrow. Combining sequence and duration on a single arrow

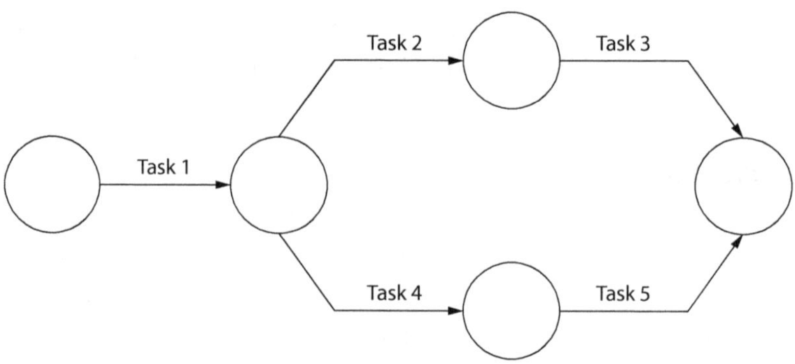

Figure 2.9 Arrow diagram.

Activity	Preceding Activity
Task 1	None
Task 2	Task 1
Task 3	Task 1
Task 4	Task 2 and Task 3

TABLE 2.4 ADM Parallel Logic Schedule

does, however, have some problems because the "i-node" and "j-nodes" of every activity must be uniquely defined when computerizing an ADM schedule. In advance of the formal discussion of ADM in the next chapter, the reader should try to develop ADM schedules for the small schedules found in Tables 2.3 and 2.4

An arrow diagram of the schedule shown in Table 2.3 will run into problems since the nodes for the "broken fences" described in the Fenced-Bar Chart discussion will have to, somehow, be reflected in the ADM schedule. Another example of a problem encountered with ADM is found in Table 2.4. When drawing the ADM schedule shown in Table 2.4, the "i-node" and "j-node" for "Task 2" and "Task 3" will not be unique. Since the "i-node" and "j-node" pairs are required to be unique in ADM, something must be added to fix ADM schedules. The resolution of this problem will be provided in Chapter 3.

Precedence Diagramming Method

John Fondahl in his 1964 report "Methods for Extending the Range of Non-Computer Critical Path Applications" brought together the results of several previous studies and developed the much of the language used for the second type of CPM diagram. In the Precedence Diagramming or Precedence Diagram Method (PDM), every activity is defined by a single "node" number. For "activity on the node," or PDM schedules, the sequence or precedence of activities is illustrated by arrows.

Figure 2.10 shows the schedule from Table 2.1 in PDM format. All of the activities' names are shown inside a box in the activity node. Unlike the ADM example in Fig. 2.9, the arrows

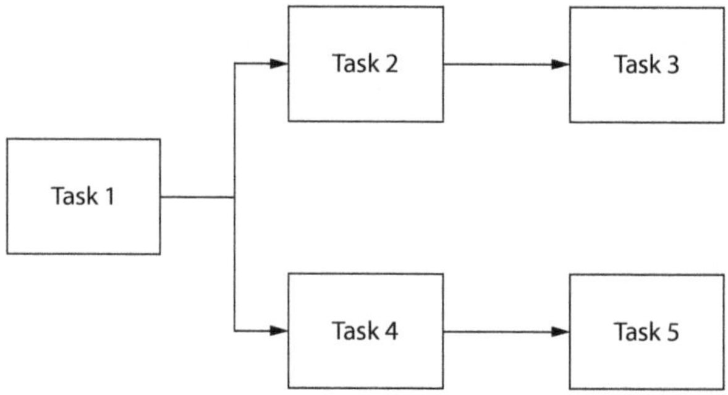

FIGURE 2.10 Precedence diagram.

in PDM are only used to show sequence. PDM schedules are typically not time-scaled. PDM splits the representation of an activity's duration from the representation of the sequence between activities. Creating PDM diagrams of Tables 2.3 and 2.4 and comparing the results to the same ADM schedules provides an intuition of the clarity of PDM. More detail will be provided about PDM in Chapter 4.

Earned Value Management

In the latter third of the 20th century, schedulers began to include more than sequence and duration on their schedules. The United States Department of Defense began to require that costs be added to CPM using a set of techniques known collectively as "Earned Value Management." This very exhaustive and detailed cost control technique is used to track costs on decades-long projects leading to the production of next-generation aircraft and ships.

For construction projects, only a small subset of Earned Value Management techniques is typically required. This is because most construction contracts are not bid on a cost-reimbursable basis. The subset of Earned Value Management typically used during construction helps to track the expected contractor billings. When the schedule is first constructed, an appropriate proportion of total construction contract amount is assigned to each activity. The assigned amount that is properly called "planned earned value" is often incorrectly referred to in construction

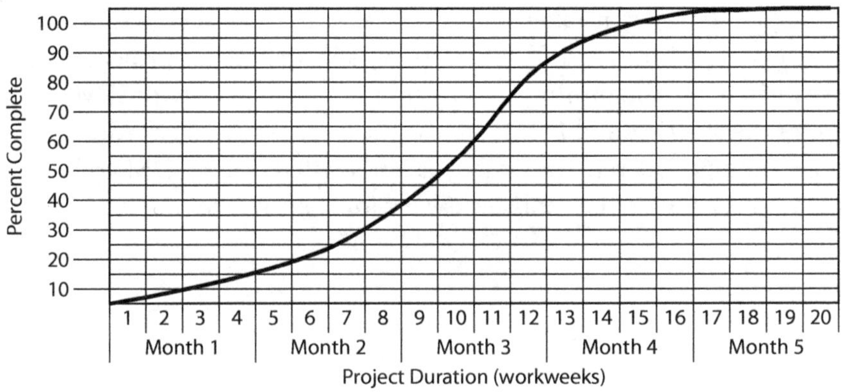

FIGURE 2.11 Earned value S-curve.

contracts as "activity cost." As the project proceeds, work on individual activities begins. The "actual earned value" is often expressed as a "percent complete" of planned earned value. The sum of all activities' actual value earned during any given month is the basis for progress payments on most construction projects today.

The typical expected earned value for most construction projects begins slowly as the contractor mobilizes the resources needed to begin construction. During the middle of the project, the majority of the earnings occur. At the end of a project, the rate of earnings decreases as the majority of crews will have moved off the job. The earned value curve has a standard shape, called an "S-curve," as shown in Fig. 2.11.

Sequence, Duration, and Progress

By themselves, humans are not the strongest of animals. They do not have heavy fur, great strength, or strong claws. However, humans do have the unique ability to plan, organize, and execute a plan to achieve complex goals. From the very start of human history, we have records of people organizing to manage great construction works.

Regardless of the planning method used during the ages, the need for scheduling has remained the same through the ages. Scheduling has three essential components: to identify a set of activities that, when taken in sequence, will achieve a goal; to estimate the duration of each activity; and to track a project's execution from beginning through to the end. Today, two CPM techniques, ADM and PDM, are the most widely used scheduling methods. The remainder of this book will discuss ADM and PDM and provide the tools necessary for their mastery.

CHAPTER 3

Arrow Diagram Method

The Critical Path Method (CPM) models construction projects by including information about sequence and duration. There are two versions of CPM. These are the Arrow Diagram Method (ADM) and the Precedence Diagram Method (PDM). In ADM, each activity is shown as an arrow. Activity arrows begin and end at nodes. ADM sequence is shown by the set of incoming and outgoing activities at each node. In PDM, each activity is shown as a node. Arrows in PDM illustrate the sequence between any two activities. Regardless of which CPM version is used, the end goal is the same: to identify a set of activities that when taken together, determine the overall length of the project. This set of activities is called the Critical Path.

Determining the Critical Path begins by defining the activities in the schedule, then defining their sequence, and finally using the duration of each activity, in turn, to determine the longest path(s) through the schedule. The paragraphs below provide the specifics of the ADM.

Activities

Once the list of activities that define the project is developed, the ADM process begins by identifying the nodes from which the activities start and stop. Figure 3.1 shows two activity arrows for "Task 1" and "Task 2" in ADM. In Fig. 3.1, "Task 1" precedes "Task 2." As the diagram is developed, each activity node

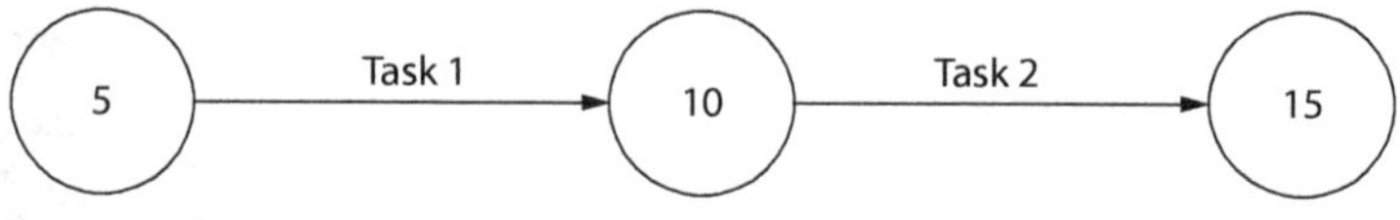

FIGURE 3.1 ADM diagram with two activities.

is uniquely numbered. In Fig. 3.1, three nodes are numbered 5, 10, and 15. ADM nodes are numbered in ascending order. As a result, the number of the starting node will always be smaller than the number of the ending node.

The starting node of a given activity is called its "i-node." The i-node of "Task 1" is 5. The ending node of a given activity is called its "j-node." The j-node of "Task 1" is 10. For "Task 2" the i-node, j-node pair is (10, 15). If the j-node of an activity matches the i-node of another activity, then the first activity can be said to precede the second activity. Given that the j-node of "Task 1," node 10, is also the i-node of "Task 2," a computer program implementing ADM will properly interpret that "Task 1" precedes "Task 2."

Precision about ADM node numbers is important for two reasons. First, software to calculate CPM schedules only does what it is told, so if the i-node and j-node numbers used to calculate the schedule are incorrect, the resulting schedule calculation will be incorrect. The second reason that precision of node numbers is important is that the relative values of the nodes themselves may be used to detect certain kinds of logic errors. If there is an activity with an i-node larger than its j-node, that error can be identified and corrected.

Ensuring that an i-node is smaller than a j-node is essential to the very notion of CPM. This is because CPM schedules are project-based, or acyclic. The most basic definition of a project is any endeavor that has a beginning and an ending. A cycle, on the other hand, loops over and over again to perform a specific operation, or set of operations. A situation where an i-node is larger than a j-node may create a loop or cycle in the schedule. To ensure that schedules conform to the basic definition of a project and to ensure proper CPM calculation, every schedule should have a single starting and a single ending node. Loops, or cycles, are not permitted in CPM.

Sequence

To begin the development of an ADM diagram, a list of activities and their sequence is needed. The first example to be used will be the set of activities previously shown in Table 2.1. The associated Fenced-Bar Chart for that schedule is shown in Fig. 2.3. The ADM diagram resulting from the schedule in Table 2.1 is shown in Fig. 3.2.

The ADM diagram in Fig. 3.2 begins with one starting node. The starting node number is 5. From this starting node, the first activity "Task 1" is added to the diagram. Node 5 is the i-node of "Task 1." The ending node of "Task 1" is set to be node 10. Node 10 is the j-node of "Task 1." According to Table 2.1, after "Task 1" activities, "Task 2" and "Task 4" may begin. To show this sequence in ADM, the j-node of "Task 1" becomes the i-node of "Task 2" and "Task 4." The activity arrows for "Task 2" and "Task 4" begin at node 10. According to Table 2.1, each of these activities has a different set of followers, therefore separate nodes are added to the diagram at the end of "Task 2" and "Task 4." "Task 3" follows "Task 2" so the j-node for "Task 2" is the i-node for "Task 3." "Task 5" follows "Task 4" so the j-node for "Task 4" is the i-node for "Task 5." To complete the project, a single ending node for both "Task 4" and "Task 5" is added to the diagram and given the next ascending node number in the series, 25.

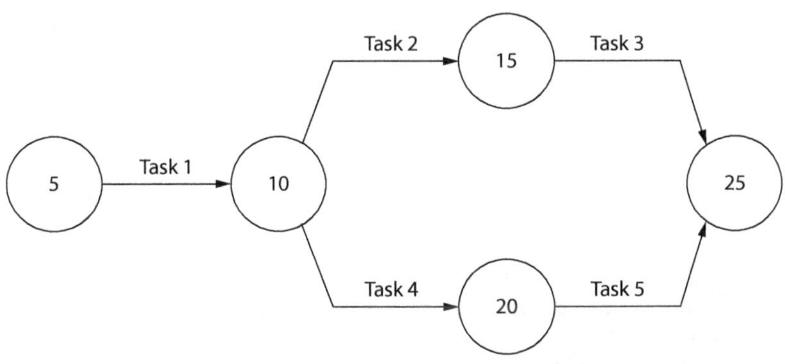

FIGURE 3.2 ADM diagram with five activities.

Numbering Dummies

Given that each pair of i-node and j-node numbers in an ADM diagram must be unique, problems arise when scheduled activities have the same sequence. Since such activities would need to begin and end at the same nodes, their pairs of starting and ending nodes would not be unique. The schedule listed in Table 2.2 is an example of a set of activities that share the same sequence. Figure 3.3 illustrates this node numbering error. In Fig. 3.3, "Task 1" and "Task 2" both incorrectly share the node pair of (5, 10). "Task 3" and "Task 4" also do not have unique node numbers.

To correct node numbering problems, ADM introduces a new type of activity called a "Numbering Dummy." The Numbering Dummy is zero duration activity added to resolve nonunique node numbering problems. Numbering Dummies are drawn on a schedule with a dashed line to allow quick visual identification. The names for these activities in a computer program are simply set to the word "Dummy."

The proper introduction of Numbering Dummies for Fig. 3.3 is shown in Fig. 3.4. Each of the two instances of nonunique node numbers, "Task 1" and "Task 2", and "Task 3" and "Task 4" have now been resolved. Two Numbering Dummies have been introduced. The first Numbering Dummy, with node numbers (10, 15), makes the node number of "Task 2" differ from those of "Task 1." The second Numbering Dummy, with node numbers (20, 25), makes the node numbers of "Task 4" differ from those of "Task 3."

A clue about when to add Numbering Dummies can be found in the activity list. Numbering Dummies occur when activities share

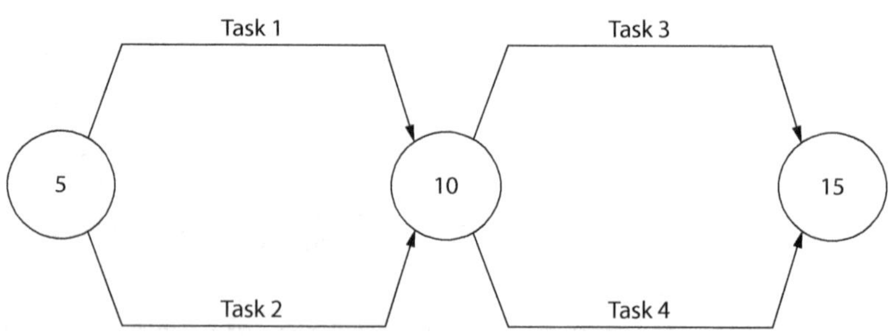

FIGURE 3.3 ADM node numbering error.

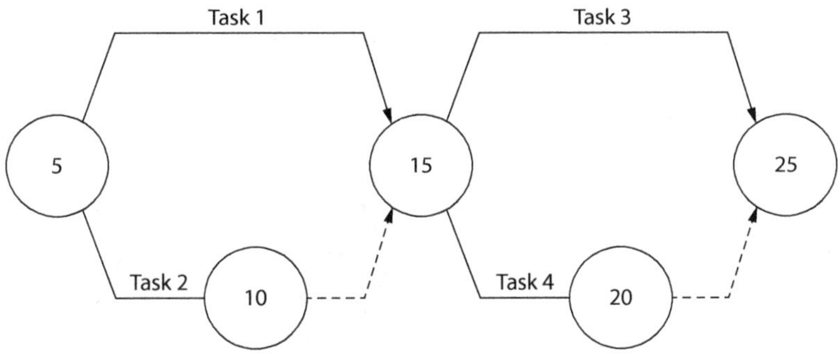

Figure 3.4 ADM with Numbering Dummies.

exactly the same set of priors and followers. Numbering dummies are then introduced to ensure unique i-node, j-node pairs.

Logic Dummies

Numbering Dummies are not the only type of dummy activity required in ADM. Logic Dummies, the second set of dummy activities required by ADM, occur when activities share some but not all of the following set of activities. Close inspection of the activity list will provide the clue that a Logic Dummy is needed.

An example of schedule with overlapping logic was presented in Table 2.3 and illustrated by a Fenced-Bar Chart in Fig. 2.5. This schedule has two starting activities, "Task 1" and "Task 2." "Task 1" and "Task 2" have some followers that are shared and some that are not shared. As a result, Logic Dummy activities are required. The Logic Dummies allow both "Task 1" and "Task 2" to affect "Task 4" without also affecting "Task 3" and "Task 5." Figure 3.5 shows how logic dummies are included in this ADM diagram. The j-node of "Task 1," node 15, is added to allow "Task 3" to follow "Task 1" without being incorrectly impeded by "Task 2," as well. The Logic Dummy on the arrow between nodes 15 and 20 also allows "Task 1" to precede "Task 4." For "Task 2," a Logic Dummy is required between nodes 10 and 20. This allows "Task 2" to precede "Task 4" and "Task 5," without also constraining "Task 3."

Logic Dummies are one of the most confusing aspects of ADM; however, they are essential for the proper representation

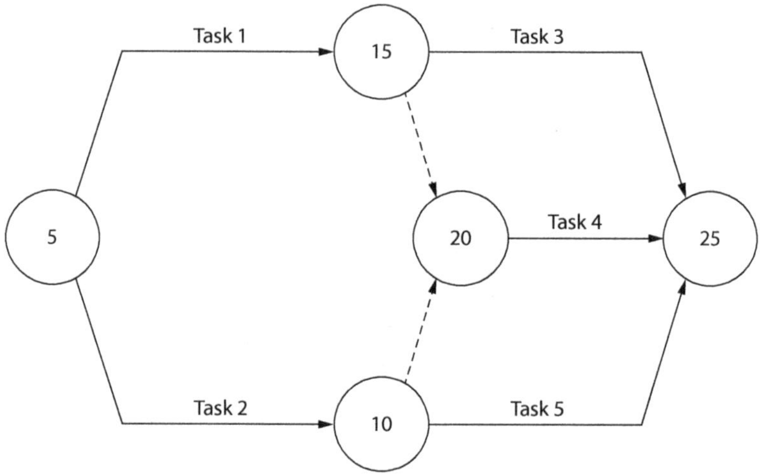

FIGURE 3.5 ADM with Logic Dummies.

of logic in cases where overlapping logic occurs in the list of schedule activities. Table 3.1 provides another example where Logic Dummies are needed. The clue that Logic Dummies are needed is that the set of priors for "Task 4," "Task 5," and "Task 6" are only partially shared. Since all three activities have "Task 2" as a prior but do not also share precedence of "Task 4" and "Task 6," one or more Logic Dummy could be expected.

A scheduler new to ADM may attempt to draw the schedule from Table 3.1 without Logic Dummies. An example of what that might look like is shown in Fig. 3.6. To see where the diagram

Activity	Duration (Workweeks)	Preceding Activity
Task 1	3	None
Task 2	5	None
Task 3	2	None
Task 4	3	Task 1 and Task 2
Task 5	1	Task 2
Task 6	2	Task 2 and Task 3

TABLE 3.1 Schedule Requiring ADM Logic Dummies

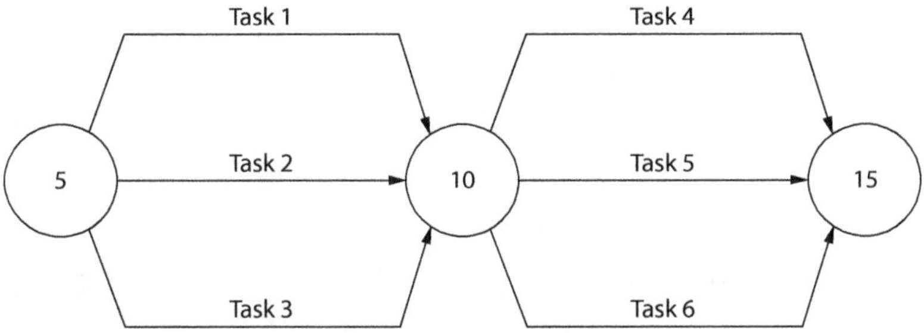

FIGURE 3.6 Incorrect ADM without Logic Dummies.

in Fig. 3.6 fails, begin by looking at each activity and then finding its preceding, or prior, activities in Table 3.1. There is no problem with "Task 1" since it has no prior activity listed in Table 3.1. Likewise, there is no problem with "Task 2" or "Task 3" since they also have no prior activities listed in Table 3.1. Given that all "Task 1," "Task 2," and "Task 3" are the first activities in the schedule, there should be a single node that starts the project. This is correctly shown in Fig. 3.6, as node 5.

The next activity in the schedule is "Task 4." To determine the prior activities of "Task 4" from the ADM diagram, begin by finding the i-node of "Task 4." The i-node of "Task 4" is identified in Fig. 3.6 as node 10. In ADM, the formal definition of a prior is any activity whose i-node is the j-node of another activity. By inspecting Fig. 3.6, it can be seen that the prior activities of "Task 4" are "Task 1," "Task 2," and "Task 3." Comparing this result with the required sequence in Table 3.1, Fig. 3.6 is shown to be incorrect. A similar pattern of incorrect logic in Fig. 3.6 can be found with "Task 5" and "Task 6."

Given that the activities in Table 3.1 have an overlapping set of sequence, Logic Dummies will be needed. Figure 3.7 provides the corrected schedule. Two Logic Dummies were inserted into the schedule. Because of these two new activities, the sequence between the activities in Table 3.1 can be correctly maintained. In Fig. 3.7, notice that "Task 5" has only one prior, "Task 2." "Task 4" is correctly constrained by both "Task 1" and "Task 2." Finally, "Task 6" has only the required two priors, "Task 2" and "Task 3."

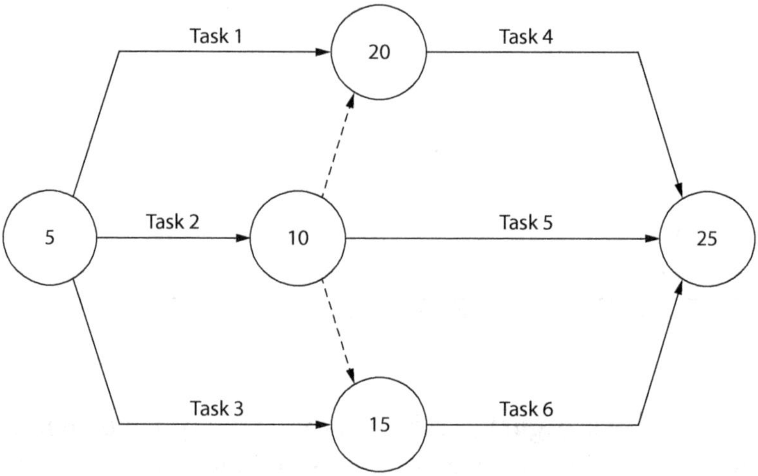

FIGURE 3.7 Corrected ADM with Logic Dummies.

Duration

Once an ADM diagram is constructed, the schedule can be calculated using the CPM process. Figure 3.8 provides the ADM diagram based on the schedule found in Table 2.1. Above each activity arrow is the name of the activity. Below each activity arrow is the estimated duration of that activity. For example, the duration of "Task 2" is 5 working days.

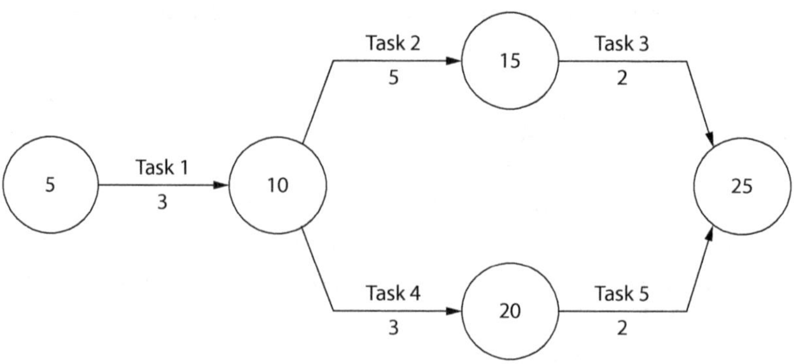

FIGURE 3.8 ADM with activity durations.

The CPM process for calculating a schedule has three distinct steps. The first step, the “Forward Pass,” answers the question “What is the earliest that an activity may begin or end?” The second step, the “Backward Pass,” answers the question “What is the latest that an activity may begin or end?” The third step, “Total Float,” answers the question “What will happen if an activity is delayed?” The sections that follow present the CPM calculation method for ADM diagrams.

Forward Pass

The Forward Pass begins at the first node of the schedule and follows through to the last node in the schedule. This “left-to-right” process of the procedure follows the notion of forward progress. Before starting the Forward Pass, two additional steps are needed. First, the ADM diagram must be updated to include “Early Event Time” boxes above each activity node. The updated ADM diagram containing these Early Event Time boxes is provided in Fig. 3.9.

The second step that is needed to calculate the Forward Pass is the date on which the project is to start. On manually calculated schedules, it is typically assumed that the work will start as soon as possible. When doing computer calculations, one will also want to know about the workweeks and holiday calendar settings for his/her crews. More will be said about this

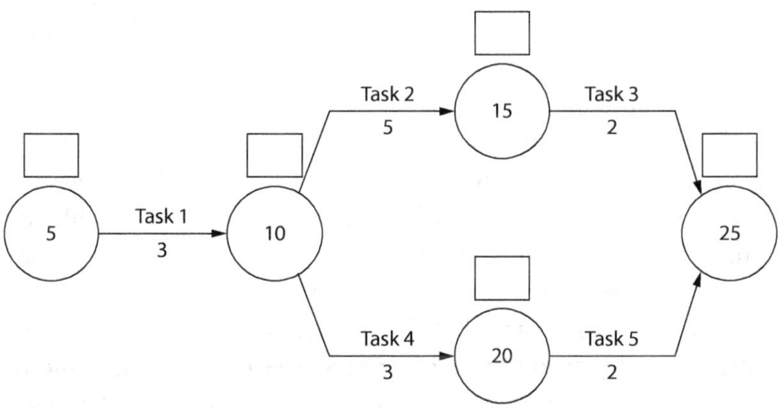

Figure 3.9 ADM with Early Event Time boxes.

topic in a later chapter. For now, begin the Forward Pass calculation by putting the number zero in the Early Event Time for the first activity in the schedule. That Early Event Time box in Fig. 3.9 is the box above node number 5.

Now that the start of the project has been set, the arithmetic of the Forward Pass can begin. For a given activity, the Early Event Time at the end of that activity is equal to the Early Event Time at the start of that activity, plus the activity's duration. For "Task 1," the Early Event Time at the end of "Task 1" is the Early Event Time at the start of "Task 1," zero, plus the duration of "Task 1," which is shown under the "Task 1" activity arrow, three. As a result, the Early Event Time of "Task 1" is three. The value of three is placed in the Early Event Time above node 10.

The Forward Pass proceeds from left to right across the ADM diagram. Applying the same procedure to the Early Event Time for "Task 2" can be seen to be the Early Event Time at the start of "Task 2," three, plus the duration of "Task 2," five. The resulting Early Event Time at the end of "Task 2" is, therefore, set to eight. The same procedure is applied to "Task 4" yielding an Early Event Time at the end of "Task 4" of six.

The Early Event Time at the end of the project, at node 25, is a special case. At node 25, there are two activities, "Task 3" and "Task 5," that enter this node. This means that there are multiple possible values for the Early Event Time at node 25. Along one path, through "Task 3," one possible value of the Early Event Time at node 25 would be 10. Along the other path, through "Task 5," the other possible value of the Early Event Time at node 25 would be eight.

To determine which of the two possible Early Event Times to select, one only has to remember that the purpose of the Forward Pass is to determine the earliest time that a project could be completed. The example, shown in Fig. 3.9, shows two alternatives. Both alternatives must be met to complete the work prior to node 25. This means that the greater of the two Early Event Times, day 10, needs to be entered as the Early Event Time of node 25. The general rule is that the Early Event Time at a given node is equal to the greatest of the incoming Early Event Times.

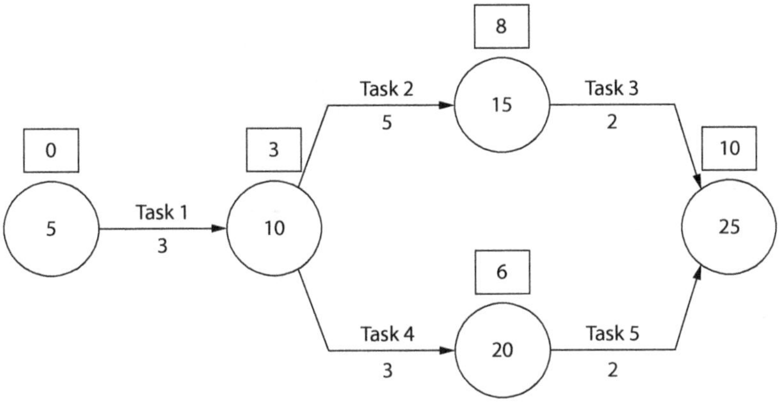

FIGURE 3.10 ADM with completed Forward Pass.

As a result of calculating the Forward Pass for the example project, it can be seen, in Fig. 3.10, that the earliest the example project can be completed is day 10.

If present in the schedule, Numbering Dummies and Logic Dummies are also included in the Forward Pass calculation. Dummy activities always have a zero day duration. As a result, they simply pass through the Early Event Times created by the real activities in the schedule.

Backward Pass

The Backward Pass begins at the last node of the schedule and follows through to the first node in the schedule. This "right-to-left" process of the procedure follows the notion of looking backward through the project. Before starting the Backward Pass, two steps are needed. First, the ADM diagram must be updated to include "Late Event Time" boxes below each activity node. The updated ADM diagram containing these Late Event Time boxes is provided in Fig. 3.11.

The second step needed to calculate the Backward Pass is the date on which the project is to end. On manually calculated schedules, it is typically assumed that the work will end as soon as possible. When doing computer calculations, one will also want to know the required contract completion date. For now, begin the Backward Pass calculation by finding the

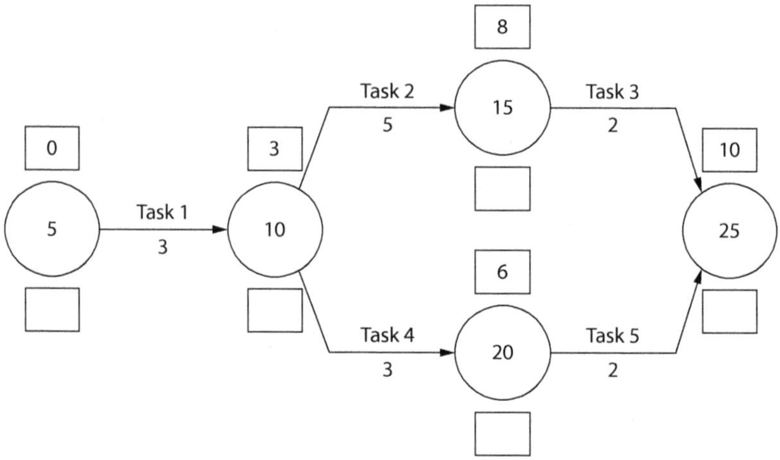

FIGURE 3.11 ADM with Late Event Time boxes.

last Early Event Time box for the last node in the schedule. At node 25, the Early Event Time is on day 10. Assuming that the project should be completed as soon as possible, place the Early Event Time at the end of the project as the Late Event Time in that same node. Once accomplished, the value of 10 will be found in the Late Event Time of node 25.

Now that the end of the project has been set, the arithmetic of the Backward Pass can begin. For a given activity, the Late Event Time at the start of that activity is equal to the Late Event Time at the end of that activity minus the activity's duration. For "Task 5," the Late Event Time at node 20 is the Late Event Time at the end of "Task 5," 10, minus the duration of "Task 5," which is two. As a result, the Late Event Time at node 20 is eight. The value of eight is placed in the Late Event Time below node 20.

In the same way, the Late Event Time at the start of "Task 3," node 15, is the Late Event Time at end of "Task 3" less the duration of "Task 3." As a result, the Late Event Time at the beginning of "Task 3" is eight. The value of eight is placed in the Late Event Time below node 15.

The Late Event Time at the start of "Task 2" and "Task 4" is a special case. At node 10, two activities, "Task 2" and "Task 4" start from that node. This means that there are multiple possible values for the Late Event Time at node 10. Along one path, through "Task 2," one possible value of the Late Event Time at node 10 would be three. Along the other path, through "Task 4," the other possible value of the Late Event Time at node 10 would be five.

To determine which of the two possible Late Event Times to select, one only has to remember that the purpose of the Backward Pass is to determine the latest time that a project could be completed. The example in Fig. 3.11 shows two alternatives. Both alternatives must be met to complete the work following node 10. This means that the earlier of the two Late Event Times, day 3, needs to be entered as the Late Event Time of node 10. The general rule is that the Late Event Time at a given node is equal to the earliest of the outgoing Late Event Times.

The last step to complete the Backward Pass is to consider "Task 1." With a Late Event Time at the j-node of "Task 1," node 10, set to three, the Late Event Time at the start of "Task 1" may be simply calculated as the Late Event Time at j-node less the activity's duration. The Late Event Time at the i-node of "Task 1" is, therefore, zero.

The result of calculating the Backward Pass for the example project can be seen in Fig. 3.12. If present in the schedule, Numbering Dummies and Logic Dummies are also included in the Backward Pass calculation. Dummy activities are given zero

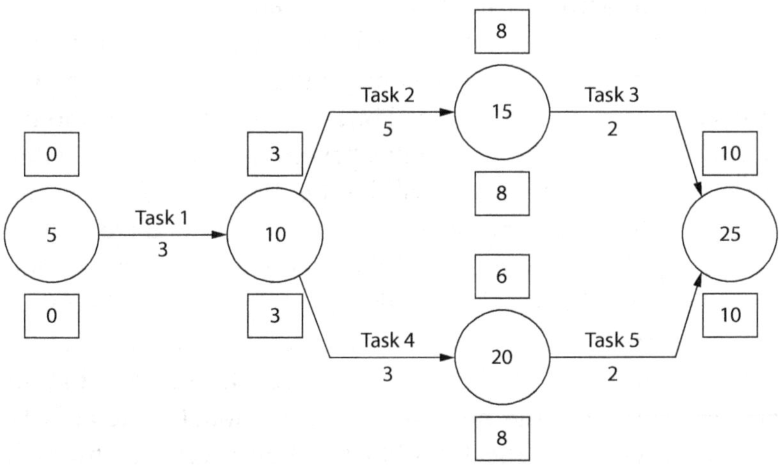

FIGURE 3.12 ADM with completed Backward Pass.

duration. As a result, they simply pass through the Late Event Times created by the real activities in the schedule.

Total Float

Most projects include activities that, if delayed, will delay the entire project. These projects will also have other activities that could be delayed days or weeks without having any appreciable impact on the overall duration of the project. The Fenced-Bar Chart in Fig. 2.4 provided an intuition about which activities could be delayed, and which could be not. The final step of CPM, Total Float Calculation, quantifies the extent to which activities may or may not be delayed.

The extent that an activity can "float" or "slip" can be seen in the following example. Consider an activity with a j-node having an Early Event Time and Late Event Time. The Early Event Time for the activity is the earliest that the activity could be completed for the project to complete on time. The Late Event Time is the latest that the activity could be complete without delaying the project completion. When comparing these values for a set of activities in series, there are three possibilities.

The first possibility is that the Early Event Time and Late Event Time values are equal. In this case, any delay in reaching that node will delay the project. The earliest that the activity can reach that node is at the end of its planned duration. The difference between the two values will be zero. The value of the Total Float for this activity will, therefore, be zero.

The second possibility is that the Late Event Times are greater than the Early Event Time. In this case, an activity may reach a node at its Early Event Time, or even later. The amount of time that the activity can take, beyond its Early Event Time without delaying the node, will be the difference between the Late Event Time and the Early Event time. This Total Float value will always be positive. As will be seen later, a positive Total Float provides the scheduler with some important flexibility.

The third possibility is that the Late Event Time is earlier than the Early Event Time. In this case, the earliest that an activity can reach the node is not early enough. The activity is behind schedule. The Total Float will be negative. Although some schedulers manipulate schedules to never obtain a negative Total Float, experienced schedulers will want to get unambiguous feedback about "late" activities. As will be seen later, resolving a negative Total Float may not be as simple as working harder on one activity to make up the time.

For a complex ADM diagram, where several activities are entering a given node, the calculation of Total Float is not quite as simple as subtracting an activity's j-node Early Event Time from the j-node Late Event Time. This is because the Total Float of each activity that ends at a given node depends on that activity's duration. The complete formula for calculating the Total Float for an ADM diagram is the that Total Float for a given activity will be equal to the value of its Late Event Time (at its j-node) less the sum of the activities' Early Event Time (at its i-node) plus the activity duration.

To begin the Total Float Calculation, the ADM diagram is updated with symbols placed next to the activity duration. In the case of Fig. 3.13, triangles are placed on the ADM diagram. These triangles will contain the results of the Total Float Calculation.

Once the symbols have been added to the ADM diagram, then the Total Float may be calculated. There is no particular directionality needed to follow when performing the Total Float Calculation; however, to ensure that all activities' Total Floats are calculated, it is often helpful simply to traverse the schedule starting at the first activity and moving through to the end of the schedule.

To begin calculating the Total Float for the schedule in Fig. 3.13, begin with "Task 1." The Late Event Time at the j-node

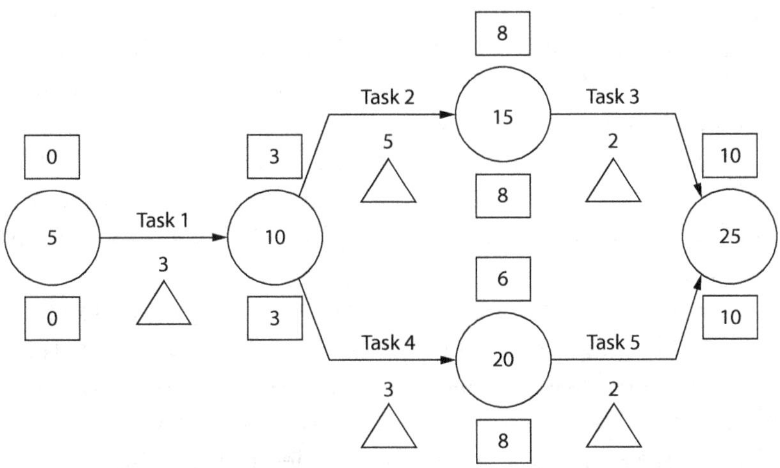

Figure 3.13 ADM with Total Float symbols.

of "Task 1" has a value of three. From that value are subtracted the sum of the Early Event Time at the i-node of the activity and the duration of the activity. The Total Float of "Task 1" is then set to zero and the number is placed within the triangle beneath the "Task 1" activity arrow.

The Total Float for "Task 2" is calculated by first finding the Late Event Time at its j-node, eight. From that value are subtracted the sum of the Early Event Time at the activity's i-node and the duration of the activity. This also yields a zero Total Float value for "Task 2." The Total Float for "Task 3" can now be calculated. And as with "Task 2" the Total Float for "Task 3" is also zero.

By inspection, the pattern of the previous activities appears different when looking at "Task 4" and "Task 5." The Late Event Time at node 20 does not match the Early Event Time at that node. Scheduling intuition indicates that there is some amount of Total Float in the activities that meet at such a node. To turn this intuition into a prediction of the performance of the two activities, the Total Float Calculation is completed.

The Total Float for "Task 4" is calculated by first finding the Late Event Time at its j-node, eight. From that value are subtracted the sum of the Early Event Time at the activity's i-node and the duration of the activity. The Early Event Time at the i-node of "Task 4" is 3 days. The duration of "Task 4" is three. The Total Float Calculation of eight minus six yields a Total Float of 2 days.

The Total Float for "Task 5" is calculated by first finding the Late Event Time at its j-node, 10. From that value are subtracted the sum of the Early Event Time at the activity's i-node and the duration of the activity. The Early Event Time at the i-node of "Task 5" is six. The duration of "Task 5" is 2 days. The Total Float Calculation of 10 minus 8 yields a Total Float of 2 days.

For this small schedule, we can determine by inspection that the Total Float calculation for "Task 4" and "Task 5" is correct. "Task 2" and "Task 3" together take a total of 7 days to complete, while "Task 4" and "Task 5" take only 5 combined days. The completed ADM diagram is provided in Fig. 3.14.

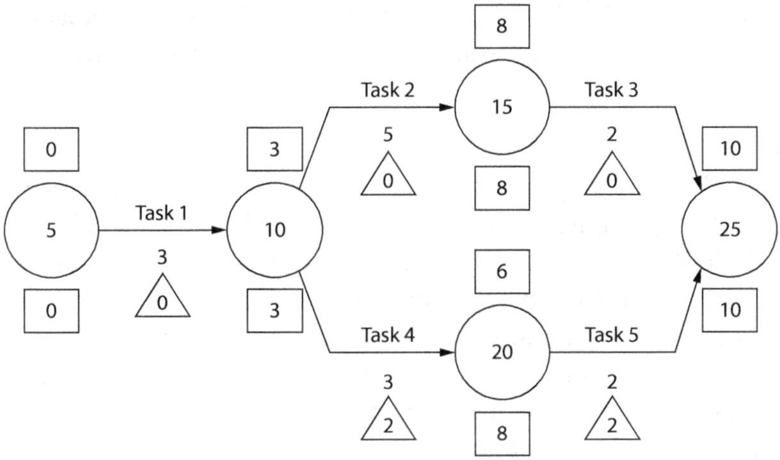

FIGURE 3.14 ADM with Total Float Calculation.

Finding the Critical Path

Now that the schedule calculations are complete, the Critical Path for the schedule can be identified. The Critical Path of a schedule is defined as that set of activities that will, if any one of the activities is delayed, result in a delay to the overall project. The mathematical way to define the Critical Path is to identify the set of activities with the least amount of Total Float.

In Fig. 3.14, the Critical Path can be identified by those activities with the minimum Total Float. That set of activities is composed of "Task 1," "Task 2," and "Task 3." The two other activities "Task 4" and "Task 5" are not on the Critical Path since they have a total float of two.

In considering the meaning of the Critical Path, it is important to recognize that the Total Float of a given path of activities is a property of the path of activities that share the Total Float. Looking at "Task 4" and "Task 5," both of these activities cannot be delayed by 2 days each. The path of "Task 4" and "Task 5" taken together may be delayed by 2 days. Consider, for example, if "Task 4" were delayed 2 days and "Task 5" were delayed 1 day. The resulting project schedule would have a new Early Event Time of 11. If the overall project completion date were required to be day 10, then a new critical path would be developed as "Task 1," "Task 4," and "Task 5." The Total Float on this new Critical Path would have a value of negative one.

Summary

ADM's underlying metaphor is one of a time line. As a result, ADM is appealing as a communication tool. Unfortunately, the simple ADM model requires a significant amount of effort to correctly identify and maintain Numbering Dummies and Logic Dummies. Furthermore, the particular nomenclature of "Event Time" scheduling may be confusing for some practitioners. While some larger owners continue to use ADM for historical purposes, its use continues to decline.

Over time, it can be expected that ADM will continue to fade into the background. The expected failure of ADM to continue beyond the 21st century can be directly attributed to modeling activity sequence and duration into a single activity arrow. The version of CPM introduced in Chapter 4 does not have such shortcomings.

CHAPTER 4

Precedence Diagram Method

The Critical Path Method (CPM) models construction projects by including information about sequence and duration. There are two versions of CPM. These are the Arrow Diagram Method (ADM) and the Precedence Diagram Method (PDM). In ADM, each activity is shown as an arrow. Activity arrows begin and end at nodes. ADM sequence is shown by the set of incoming and outgoing activities at each node. In PDM, each activity is shown as a node. Arrows in PDM illustrate the sequence between any two activities. Regardless of which CPM version is used, the end goal is the same: to identify set of activities that, when taken together, determine the overall length of the project. This set of activities is called the Critical Path.

Determining the Critical Path begins by defining the activities in the schedule, then defining their sequence, and finally using the duration of each activity, in turn, to determine the longest path(s) through the schedule. The paragraphs below provide the specifics of PDM.

Activities

Once the list of activities that define the project is developed, the PDM process begins simply by listing each activity in the schedule. Unlike ADM, where each activity requires two numbers, the i-node and j-node, PDM requires only a single activity number for each activity. There is one constraint, imposed by convention, on PDM activity numbering. That convention is to ensure that activities are numbered in ascending order from the beginning of the project to the end. While not technically required, this convention helps to ensure that loops are not included in the schedule by mistake. Loops are not allowed in CPM since it models projects—which have a defined start and end.

Sequence

On ADM diagrams, the activity arrow represents both the activity and its sequence. Modeling of both sequence and duration with one symbol, the activity arrow, led directly to the need for Numbering Dummies and Logic Dummies. In PDM diagrams, the activity node simply represents the activity. The sequence among activities is represented by a sequence arrow. The sequence between every prior and every successor activity is, therefore, uniquely shown on the PDM diagram.

To begin the development of a PDM diagram, a list of activities and their sequence is needed. The first example to be used will be the set of activities previously shown in Table 2.1. The associated Fenced-Bar Chart for that schedule is shown in Fig. 2.3. The PDM diagram resulting from the schedule in Table 2.1 is shown in Fig. 4.1.

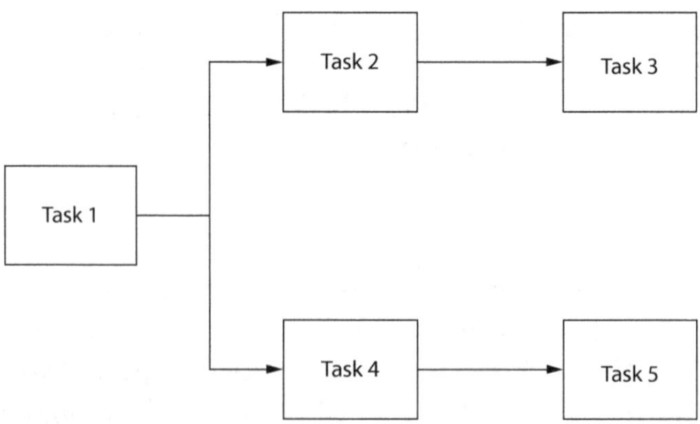

Figure 4.1 PDM diagram with five activities.

The PDM diagram in Fig. 4.1 begins with a single starting activity, "Task 1." Following "Task 1" are two activities, "Task 2" and "Task 4." To show this sequence, an arrow is placed at the right, or "finishing" side of "Task 1" and drawn to the left or "starting" side of "Task 2." To show the additional sequence between "Task 1" and "Task 4," another sequence arrow is placed between the "finish" of "Task 1" and the "start" of "Task 4." The activities that follow "Task 2" and "Task 4" are then added to the drawing and the respective "Finish-to-Start" arrows are included.

The schedule listed in Table 2.2 is an example of a set of activities that share the same sequence. In ADM, this schedule requires Numbering Dummies, as shown in Fig. 3.4. This same schedule in PDM is drawn as shown in Fig. 4.2. There are two followers of "Task 1," "Task 3" and "Task 4." The same set of followers occurs for "Task 2." Each individual relationship is defined by a separate sequence arrow.

An example schedule with overlapping logic was presented in Table 2.3, and illustrated by a Fenced-Bar Chart in Fig. 2.5. In ADM, this schedule requires Logic Dummies, as shown in Fig. 3.5. This same schedule is simply drawn in PDM as shown in Fig. 4.3.

In Fig. 4.3, "Task 1" precedes "Task 3" and "Task 4." To show this sequence, an arrow is drawn from the right, or "finishing" side, of "Task 1" to the left, or "starting" side, of "Task 3." A similar arrow is drawn from the finish of "Task 1" to the start of "Task 4." "Task 2" precedes "Task 4" and "Task 5." One arrow is

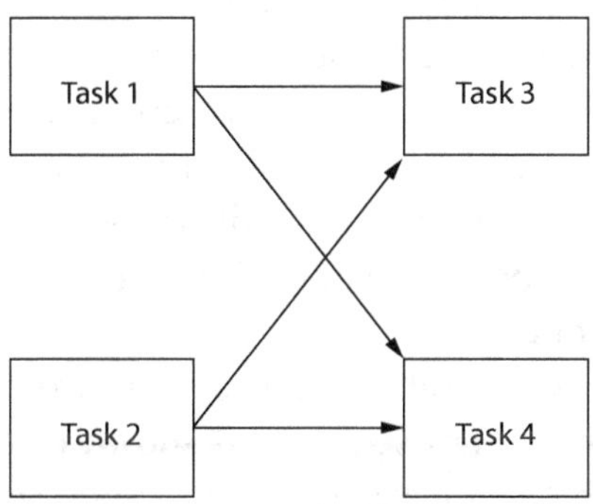

FIGURE 4.2 PDM diagram with shared sequence.

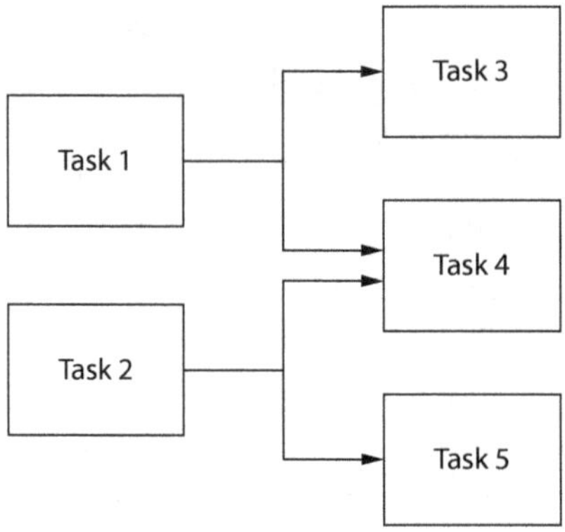

FIGURE 4.3 PDM diagram with overlapping sequence.

drawn from the finish of "Task 2" to the start of "Task 4." Another arrow is drawn from the finish of "Task 2" to the start of "Task 5"

Finish-to-Start

In the PDM diagrams above, the sequence between activities where one activity finishes before the next activity starts is described. This traditional sequence is called a "Finish-to-Start" sequence. Because PDM uses activity arrows to only show sequence, Finish-to-Start is only one of several types of sequences that can be described in PDM. Two additional types of PDM sequence are described below.

Start-to-Start

Another type of sequence described by PDM is one where the start of an activity is contingent upon the start of a preceding activity. This is the "Start-to-Start" sequence. Such overlapping sequence can be seen in many construction settings. For example, if one crew is installing drywall and another crew is following that activity to tape and finish the drywall, then the activity "Install Drywall" and "Tape and Finish" may be said to have a Start-to-Start sequence.

In Fig. 4.4, two activities are shown with a Start-to-Start sequence. A Start-to-Start sequence is shown by an arrow drawn from the bottom-left, or starting side, of the preceding activity. Depending on the position of the following activity on the diagram,

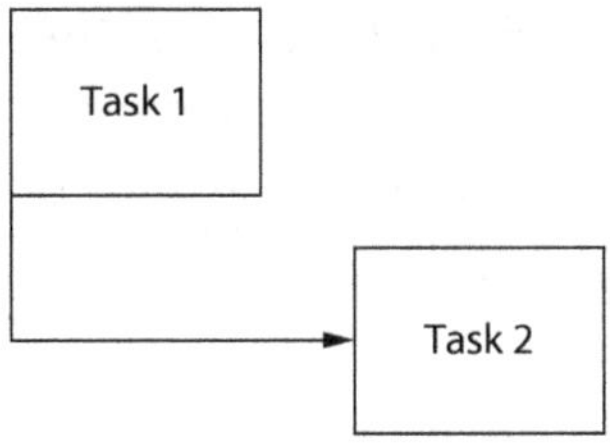

FIGURE 4.4 PDM diagram with Start-to-Start.

the tail of the sequence arrow could also be drawn from the top-left of the preceding activity. The Start-to-Start sequence arrow is then connected with the left, or starting side, of the following activity.

Finish-to-Finish

Another type of sequence described by PDM is one where the finish of an activity is contingent upon the finish of a preceding activity. This is the "Finish-to-Finish" sequence. The Finish-to-Finish sequence is typically used in combination with the Start-to-Start sequence to model activities that follow one another. As with the previous example, one crew is working on an activity "Install Drywall" and another crew is following to complete the "Tape and Finish" activity. It would be expected that shortly after completing the installation of drywall, the taping and finishing would also be complete. This logic can be represented by the Finish-to-Finish sequence.

In Fig. 4.5, two activities are shown with a Finish-to-Finish sequence. A Finish-to-Finish sequence is shown by an arrow drawn from the right, or finishing side, of the preceding activity. The Finish-to-Finish sequence arrow is then connected to the

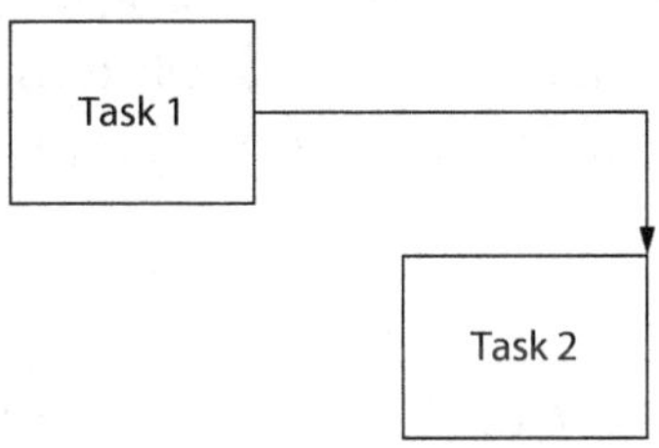

FIGURE 4.5 PDM diagram with Finish-to-Finish sequence.

top-right, or finishing side, of the following activity. Depending on the position of the following activity on the diagram, the head of the sequence arrow could also be drawn to the bottom-right of the following activity.

Lead and Lag

An interesting side-effect of separating the representation of sequence from that of duration is the addition of duration to the sequence between activities. For a Start-to-Start sequence, this duration indicates the length of time after starting the preceding activity that the following activity can begin. In the example of the Start-to-Start sequence between "Install Drywall" and "Tape and Finish," "Tape and Finish" would not usually begin immediately after the drywall begins. There will often be an amount of time after the "Install Drywall" activity would start before, or "lead", the following activity.

For a Finish-to-Finish sequence, this duration indicates the length of time after finishing the preceding activity that the following activity can finish. In the example of the Start-to-Start sequence between "Install Drywall" and "Tape and Finish," "Tape and Finish" would not usually finish immediately after the drywall finishes. There will often be an amount of time that the "Tape and Finish activity" takes, or "lags," behind the prior activity.

Given the addition of sequence type, Finish-to-Start, Start-to-Start, or Finish-to-Finish, and sequence durations, tabular listings of PDM diagrams may contain more columns of information than do ADM diagrams. On these tabular listings, Finish-to-Start is often abbreviated as "FS," Start-to-Start is abbreviated "SS," and "Finish-to-Finish" is abbreviated as "FF." Following this abbreviation, Start-to-Start and Finish-to-Finish sequence should also identify the lead or lag associated with that sequence. In general, it is assumed that the "Finish-to-Start" sequence, the traditional type of sequence, has a lag of zero.

Duration

To demonstrate the CPM calculation in PDM, a PDM diagram with nodes containing the activity name, activity duration, and calculation dates must first be provided. A PDM

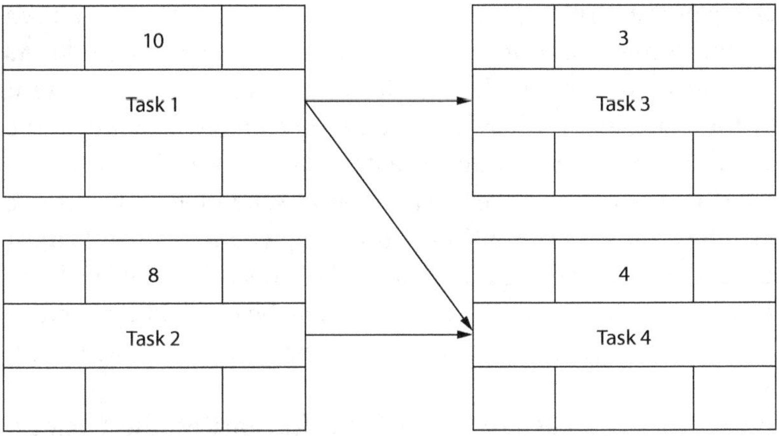

FIGURE 4.6 PDM diagram with activities and durations.

schedule used for this example is provided in Fig. 4.6. This diagram includes four activities. "Task 1" precedes "Task 3" and "Task 4." "Task 2" precedes only "Task 4." The durations of each of these tasks is provided above the task name in workdays.

The CPM process for calculating a schedule has three distinct steps. The first step, the "Forward Pass," answers the question "What is the earliest that an activity can begin or end?" The second step, the "Backward Pass," answers the question "What is the latest that an activity may begin or end?" The third step, "Total Float," answers the question "What will happen if an activity is delayed?" The sections that follow present the CPM calculation method for PDM diagrams.

Forward Pass

The Forward Pass begins at the first activity in the schedule and follows through to the last activity in the schedule. This "left-to-right" process of the Forward Pass follows the notion of forward progress. When calculating the Forward Pass for PDM, two dates are calculated for each activity: "Early Start Date" and "Early Finish Date." Depending on the specific text or software, there are different conventions for where to place these dates. In some cases, "Early Dates" are placed in the top-left and top-right boxes of the activity node. In other cases, these dates are placed on the top-left and bottom-left boxes of the activity node. As with any project scheduling effort, it is ultimately irrelevant to the calculation where these dates are placed as long as the scheduler is consistent and the convention used is communicated to the entire project team. Good schedulers will be able to determine the convention in use at a given project and adjust as needed to get their points across. For this book, the left-hand side boxes, the top-left and bottom-left boxes, will be used to hold the "Early Start Date" and "Early Finish Date" of each activity's node.

Two more steps are required before starting the Forward Pass. The first is to determine the date on which the schedule begins. For this example, the schedule will begin on day 1. The second step is to determine which activities start the schedule. In Fig. 4.6, there are two starting activities: "Task 1" and "Task 2." The Early Start Date of 1 is, therefore, placed in the upper-left box of both "Task 1" and "Task 2."

To calculate the Early Finish Date of an activity, add the activity's duration to the activity's Early Start Date. The Early Finish Date of "Task 1" is, therefore, day 1 plus 10 days of duration. The impact of calendars and holidays is covered in Chapter 5. The Early Finish Date of 11 is placed in the bottom-left box of "Task 1." The Early Finish Date of "Task 2" is equal to its Early Start Date, of 1, plus its duration, 8. As a result, the Early Finish Date of 9 is placed in the bottom-left box of "Task 2."

Next, consider the value of the Early Start Date of "Task 3." Since "Task 3" begins following the completion of "Task 1," and there are no other activities preceding "Task 3," the earliest that "Task 3" can begin is at the end of "Task 1." The Early Finish Date of "Task 1" is therefore used to set the "Early Start Date" of "Task 3," at day 11. To calculate the Early Finish Date of "Task 3," simply add the duration of "Task 3" to its Early Start Date. The Early Finish Date of "Task 3" is, therefore, day 14.

Now consider "Task 4." Notice that two activities precede "Task 4." Whenever there are more than one preceding activities, the scheduler has a choice about which incoming Early Finish Date to set as the following activities Early Start Date. The choices for "Task 4" are day 11, from "Task 1," or day 9, from "Task 2."

To determine which of the two possible Early Start Dates to select, one only has to remember that the purpose of the Forward Pass is to determine the earliest time that a project could be completed. This means that the greater of the two Early Start Dates, day 11, needs to be entered as the Early Start Date of "Task 4." The general rule is that the Early Start Date is equal to the greatest of the incoming Early Start Dates.

To complete the Forward Pass calculation, determine the Early Finish Date of "Task 4." The Early Finish Date of any activity in PDM is the sum of the Early Start Date plus the activity duration. The Early Finish Date of "Task 4" is day 15. Figure 4.7 illustrates the completed forward pass calculation.

Backward Pass

The Backward Pass begins at the last activity in the schedule and follows through to the first. This "right-to-left" process follows the notion of looking backward through the project.

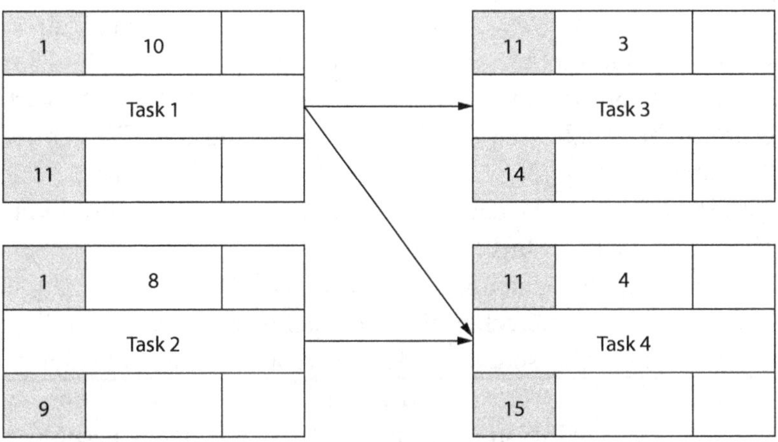

FIGURE 4.7 PDM with Completed Forward Pass.

When calculating the Backward Pass for PDM, two dates are calculated for each activity: "Late Start Date" and "Late Finish Date." Depending on the specific text or software, there are different conventions for where to place these dates. In some cases, "Late Dates" are placed in the bottom-left and bottom-right boxes of the activity node. In other cases, these dates are placed on the top-right and bottom-right boxes of the activity node. For this book, the two right-hand side boxes, will be used to hold the "Late Start Date" and "Late Finish Date" of each activity's node.

Two more steps are required before starting the Backward Pass. The first is to determine the date on which the schedule ends. The PDM diagram in Fig. 4.7 does not have a single starting and ending activity. As a result, some inference is required to determine the overall earliest that the project could complete. The example project in Fig. 4.7 has two completing activities: "Task 3" and "Task 4." The later of these activities' Early Finish Dates, day 15, is the latest date, and the earliest that the project can finish. Without additional information about the contract completion date, the latest Early Finish Date can be set as the project ending date.

The second step before starting the Backward Pass is to place that project ending date as the Late Finish Date of all the activities at the end of the schedule. In Fig. 4.7, the Late Finish Date of 15 is placed in the bottom-right box of activity "Task 3" and "Task 4."

For a given activity, the calculation of the Late Start Date is simply the subtraction of the activity's duration from the Late Finish Date of that activity. In the case of "Task 4," the Late Start Date will be equal to day 15 minus the 4-day duration. Therefore, the Late Start Date of "Task 4" is 11. In the case of "Task 3," the Late Start Date will be equal to day 15 minus the 3-day duration. Therefore, the Late Start Date of "Task 3" is 12.

Following the calculation of the Late Start Dates of the last activities in the schedule, the Backward Pass moves forward through the schedule. The next activity to consider is "Task 2." For an activity with a single prior activity, such as "Task 2," the Late Finish Date of the activity will be the Late Start Date of the following activity. The Late Finish Date of "Task 2" is, therefore, day 11. The Late Start Date of "Task 2" is the Late

Finish Date of "Task 2," day 11 minus its duration, 8 days. The Late Start Date of "Task 2" is, therefore, day 3.

The Late Finish Date at the end of "Task 1" is a special case. Two preceding activities are "Task 3" and "Task 4." This means that there are multiple possible values for the Late Finish Date of "Task 1." One possible value, from the Late Start Date of "Task 3" is day 12. The other possible value, from the Late Start Date of "Task 4" is day 11.

To determine which of the two possible Late Start Dates to select, one only has to remember that the purpose of the Backward Pass is to determine the latest time that a project could be completed. In the example project, both prior activities must have been started for "Task 1" to have finished. This means that the earlier of the two dates, day 11, needs to be entered as the Late Finish Date of "Task 1." Selecting any day later than that would delay "Task 4." The general rule is that the Late Finish Date of an activity with multiple followers is equal to the earliest of the following activity's Late Start Dates.

With the Late Finish Date of "Task 1" now established as day 11, the Backward Pass can be completed. The final calculation is to determine the Late Start Date of "Task 1." The Late Start Date of an activity is equal to its Late Finish Date, day 11, minus its duration, 10 days. The Late Start Date of "Task 1" is, therefore, day 1. The example project with the complete Backward Pass calculation is shown in Fig. 4.8.

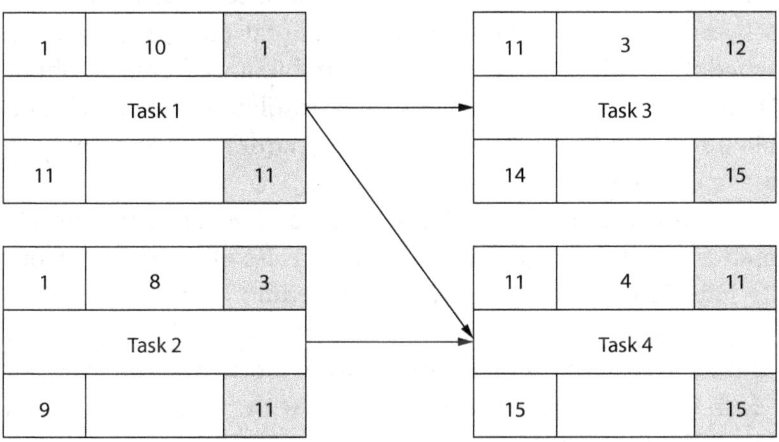

FIGURE 4.8 PDM with Completed Backward Pass.

Total Float

Total Float is the amount of time a set of activities may be delayed without delaying the project completion. Total Float is calculated for each activity by subtracting the Early Finish Date of the activity from its Late Finish Date. This situation presents three possibilities. The first is that the Late Finish Date of an activity is equal to its Early Finish Date. In this case, there will be zero Total Float. This can be interpreted to mean that if there is any delay in the completion of the activity, then the start of the following activity will be delayed, and the entire schedule will be pushed back. The second possibility is that the Late Finish Date of an activity is greater than its Early Finish Date. In this case, Total Float will be a positive value. The difference between these two dates can be seen as equal to the amount of "extra" time available to complete the activity without delaying the overall project completion. Thus, the activity completion can be said to be able to "slip" or "float" between the Early Finish Date and the Late Finish Date.

The third possibility is that the Late Finish Date is earlier than the Early Finish Date. In this case, the earliest that an activity can complete is not early enough. The Total Float will have a negative value. To get the project back "on schedule," work on that activity must be accelerated so that the activity can complete no later than its Late Finish Date. The amount that the activity must be accelerated is the duration of the activity plus that negative Total Float. Although some schedulers manipulate schedules to never obtain a negative Total Float, experienced schedulers will want to get that unambiguous feedback about "late" activities. As will be seen later, resolving a negative Total Float may not be as simple as working harder on one activity to make up the time.

As noted above, calculating the Total Float is the simple matter of subtracting the Early Finish Date of an activity from its Late Finish Date. In example schedule from Fig. 4.8, the Total Float of "Task 1" is seen to be zero. The Total Float of "Task 2" is two. The Total Float of "Task 3" is one. The Total Float of "Task 4" is zero. The example project with the complete Total Float calculation is shown in Fig. 4.9.

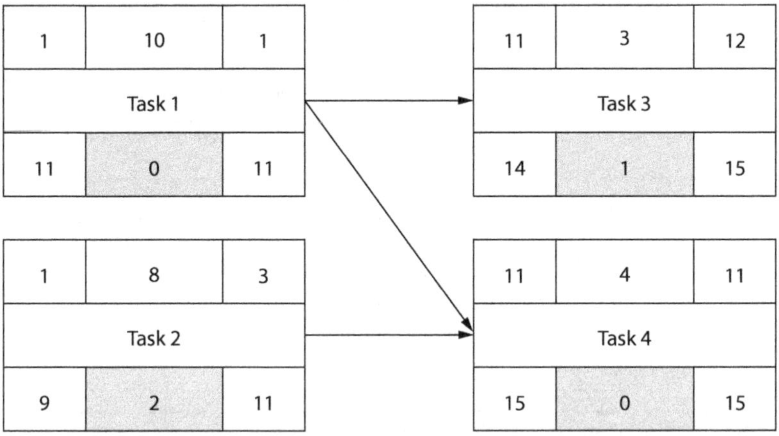

Figure 4.9 PDM with Total Float calculation.

Impact of Lead and Lag

Figure 4.9 provides a PDM schedule where the sequence between every activity is of the type Finish-to-Start. Two additional types of sequence, Start-to-Start and Finish-to-Finish, also impact CPM calculation. In the case of the Start-to-Start sequence, the sequence itself may have a "Lead" duration. This is the duration that that prior activity will start before, or lead the following activity. In the case of the Finish-to-Finish sequence, the sequence duration itself may have "Lag" duration. This is the duration that the completion of the following activity lags behind its prior activity.

Both the Start-to-Start Lead and the Finish-to-Finish Lag impact the CPM calculation. Start-to-Start Lead is only used in the Forward Pass calculation. Finish-to-Finish Lag is only used in the Backward Pass calculation.

Figure 4.10 is an updated example schedule showing new Start-to-Start logic between "Task 1" and "Task 2." To clearly distinguish this logic from the traditional Finish-to-Start sequence elsewhere in the schedule, the Start-to-Start logic is labeled as "SS." In addition to having a clear label, the duration of the sequence is also listed as 4 workdays. The meaning of this Lead duration is that "Task 2" cannot begin until four days after "Task 1" has begun.

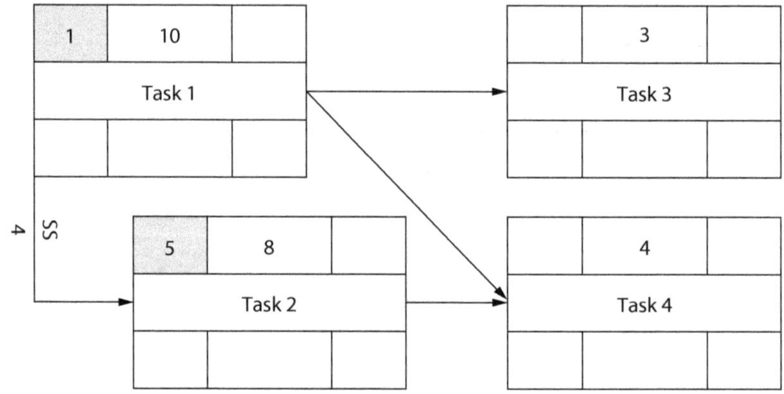

Figure 4.10 PDM with Start-to-Start Lead.

When conducting the Forward Pass calculation between two activities linked with Start-to-Start logic, the Early Start Date of the following activity is set equal to the Early Start Date of the prior activity plus the Lead duration. In Fig. 4.10, it can be seen that the Early Start Date of "Task 2" is the Early Start Date of "Task 1," day 1, plus the Lead duration of 4 days.

The example of Fig. 4.10 is very simple, with only a single activity preceding "Task 2." If there had been more than one prior activity, then the incoming Early Start Date from a Start-to-Start sequence would be one of several candidates. The selection of these candidates is governed by the multiple prior rule of the Forward Pass. This rule is that the Early Start Date of the following activity is the latest of any incoming Start Dates.

Figure 4.11 is an example schedule showing Finish-to-Finish between "Task 4" and "Task 3" and "Task 4" and "Task 5." To clearly distinguish this logic from the traditional Finish-to-Start sequence, the Finish-to-Finish logic is labeled as "FF." In addition to having a clear label, the duration of the sequence is also listed as 1 workday. The meaning of this Lag duration is that "Task 3" and "Task 5" cannot complete until 1 day before "Task 4" has completed.

When conducting the Backward Pass calculation between two activities linked with Finish-to-Finish, the Late Finish Date of the prior activity is set equal to the Late Finish Date

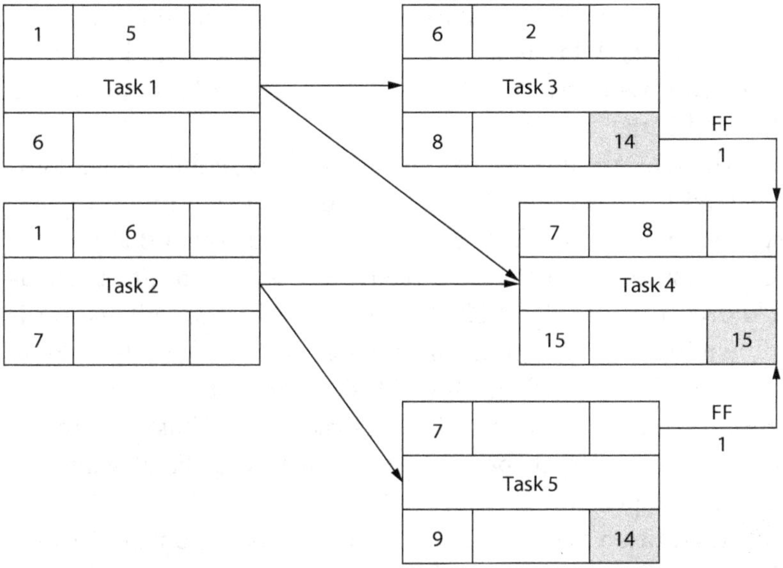

FIGURE 4.11 PDM with Finish-to-Finish Lag.

of the prior activity less than the Lag duration. In Fig. 4.11, it can be seen that the Late Finish Date of "Task 5" is 15. As a result, the Late Finish Dates of "Task 3" and "Task 5" are set to 14.

The example of Fig. 4.11 is very simple, with only a single activity following "Task 3" and "Task 5." If there had been more than one following activity, then the incoming Late Finish Date from a Finish-to-Finish sequence would be one of several candidates. The selection of these candidates is governed by the multiple follower rule of the Backward Pass. This rule is that the Late Finish Date of a prior activity is the earliest of any outgoing Finish Dates.

Finding the Critical Path

Now that the schedule calculations are complete, the Critical Path for the schedule can be identified. The Critical Path of a schedule is defined as that set of activities that will, if any one of the activities is delayed, result in a delay to the overall project. The mathematical way to define the Critical Path is to identify the set of activities with the least amount of Total Float. In Fig. 4.9 that set of activities is "Task 1" and "Task 4." The other two activities "Task 2" and "Task 3" are not on the Critical Path.

The identification of the Critical Path is slightly more complicated for schedules that have Lead and Lag. This is because it is possible for the Critical Path to travel through that Lead or Lag. First, consider the example from the Start-to-Start schedule presented in Fig. 4.10. The completed CPM calculation is shown in Fig. 4.12. The set of activities with the least Total Float is "Task 2" and "Task 4." Both "Task 2" and "Task 4" have zero Total Float. The other activities, "Task 1" and "Task 3," are not on the Critical Path because they have Total Float values of 2 and 3, respectively.

In addition to "Task 2" and "Task 4" being on the Critical Path, the Start-to-Start logic between "Task 1" and "Task 2" is also on the Critical Path. This is because any change to the Lead duration will impact the overall completion of the project. If, for example, the Start-to-Start Lead between "Task 1" and "Task 2" were only 3 days instead of 4. With this change, the new project completion is day 16. This is a 1-day decrease compared to the original project duration in Fig. 4.11. As a further example, consider the case when the Start-to-Start Lead between "Task 1" and "Task 2" is zero days. In this case, the

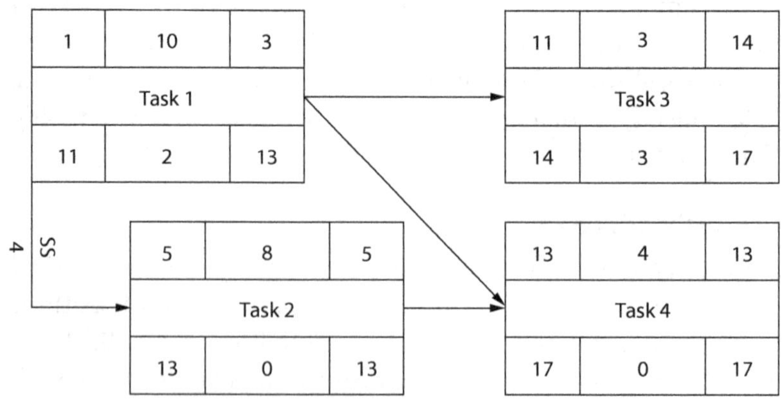

FIGURE 4.12 PDM with Start-to-Start calculation.

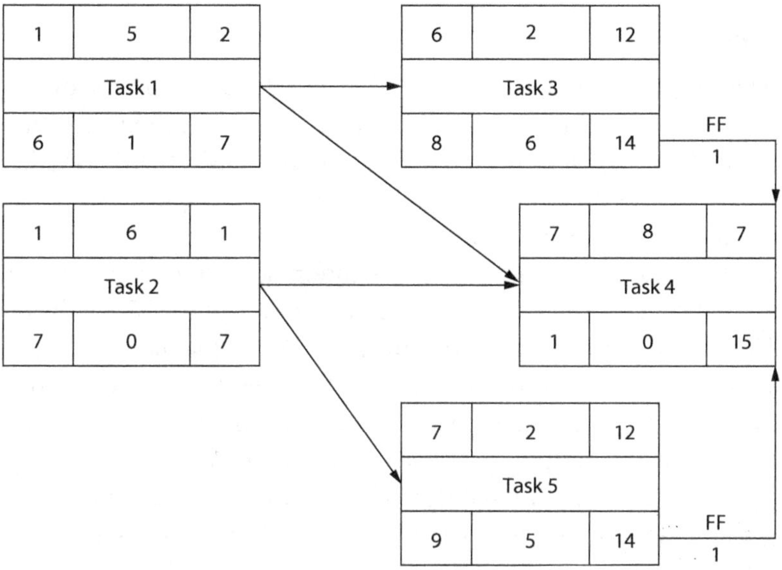

FIGURE 4.13 PDM with Finish-to-Finish calculation.

example calculation for each activity remains as shown in the original example from Fig. 4.9.

With Finish-to-Finish Lags, the Critical Path may also be governed by the Lag durations. The completed CPM calculation from the example in Fig. 4.11 is shown in Fig. 4.13. In this example, the Critical Path is a set of two activities: "Task 2" and "Task 4." In Fig. 4.11, the Lag durations are not part of the Critical Path.

The impact of Lag durations on the Critical Path could be quickly seen, if the Lag duration between "Task 4" and "Task 3" was changed to 10 days. In this case, the Late Finish Date of "Task 3" would be set to the Late Finish Date of "Task 4," day 15, minus the Lag duration of 10, which is day 5. The Late Start Date of "Task 3" would then become day 5 less the duration of "Task 3," 2 days, which is day three. Looking backward through the schedule in the Backward Pass process, the Late Finish Date of "Task 1" would then become day three and the Late Start Date of "Task 1" would be −2 days. If that Finish-to-Finish lag were a reality, then the project begins behind schedule before it has even started!

Summary

ADM uses arrows to represent both the sequence and the duration of construction activities. As a result, Numbering and Logic Dummy activities are needed in ADM. PDM, on the other hand, separates the activity from its sequence and does not need Dummy activities.

It is also interesting to note the difference between the "Early Start Date" and "Early Finish Date" in PDM, with the "Early Event Time" at the i-node and j-node of ADM. ADM calculation occurs at nodes, separate from the activities themselves. That is why the i-node and j-node boxes in ADM are called "Event Times." In PDM, specific starting and ending dates are calculated directly for each individual activity in the schedule. This distinction is another reason why PDM is more readily accepted than ADM today.

The improved representation of sequence in PDM does, however, come with its own concerns. In ADM, the only type of schedule logic allowed is Finish-to-Start. In PDM, Start-to-Start and Finish-to-Finish sequences are allowed. These new types of logic are powerful tools, allowing the scheduler to represent parallel activities more clearly. The Start-to-Start and Finish-to-Finish sequences also add Leads and Lags. These sequence durations may directly impact Critical Path calculations in unexpected ways. The scheduler, and all stakeholders, should carefully review all Lead and Lag durations to ensure that the values are reasonable.

CHAPTER 5

Creating the Plan

Readers of this book understand the qualities of effective scheduling because people plan projects every day. Consider, for example, a plan to take a cross-country trip with friends and/or family. Determining the destinations is likely to be one of the first decisions to be made. Along with this decision is that of identifying who will go on the trip and how everyone will get there. Correspondence during the planning will need to confirm the plans with each party and ensure that each member of the party has the resources needed. As the time draws close, individual members of the party will need to pack. Many people find that drawing up a packing checklist is helpful to make sure that nothing is forgotten.

Some of the items on the checklist are within the ability of the traveler, themselves, to complete. For example, the traveler may be responsible for arranging transportation and doing any needed laundry. Checklist items may, however, be contingent upon others. A flight check-in is only allowed 24-hours prior to departure. Laundry, if business attire is required, may need to be processed through a dry-cleaner. To a greater or lesser degree, people share their checklists with close traveling companions to determine if these preparatory trips can be combined.

Once the trip is under way, the traveling party will want to compare their progress with that which was expected. If there is an unexpected gate change, or road construction, the group will need to adjust their original plan. Flight delays due such as mechanical problems, or car accidents, do happen, no matter how much we plan. If the delays are caused by others, then the travelers may be able to receive compensation. Other types of delays, such as those caused by weather, are noncompensable.

If the traveling party is all on the same itinerary, determining the plan, progress, and impact of changes is as simple as having a conversation. If, however, members of the party are on different travel itineraries and they all are planning to converge for a specific event, such as a wedding celebration, then communication may be more difficult. When the parties are not all together, they will all need to know each other's itineraries and have an update of scheduled progress. Often it is not a delay itself that causes major problems for other party members—it is the lack of clear communication about the delay that causes problems for others. Airlines have certainly understood this, as evidenced by changes in airline check-in systems allowing travelers to authorize the notification of others about changes to flight status.

Most readers will have been on many such trips, either across town or across country. Common to a successful conclusion to any of such trips are the (1) unambiguous communication of the ultimate and any intermediate destinations, (2) a description of the travel itinerary, (3) an expectation of when the destinations can be reached, and (4) regular updates on the progress toward the goal.

Regardless of the size of the project or the technology used for planning, the goals of project planning remain the same. First, scheduler needs to communicate time-based project deadlines. Next, the scheduler needs to work with each trade to determine how those goals will be achieved. Based on that information, the scheduler can then predict the expected completion of the project—both on the calendar and in the bank. Finally, when changes inevitably occur, the scheduler can help the team evaluate their impact and mitigate further delays.

In the previous two chapters, the Critical Path Method (CPM) was presented as a technique to model the sequence between activities and determine how the activities' duration affects a project. The mechanics of the Arrow Diagram Method (ADM) and the Precedence Diagram Method (PDM) were introduced, and the pros and cons of each were briefly explained. The Critical Path itself was identified the set of activities with the longest path through project. This chapter shows how CPM, the arithmetic of scheduling, often-used techniques and rules-of-thumb, the art of scheduling, can be brought together to create an effective project plan.

Milestones

When the speed of travel was limited to that provided by feet and horses, stone markers, or milestones, were used to mark the distance along the road. E. H. Warmington's 1940 English translation of the "Corpus Inscriptionum Latinarum," shows that Roman milestones identified the distance, as measured in 1,000 paces (what we now call a mile), to the next city and to Rome. Thinking about how a Roman citizen might travel on the Appian Way, periodically sending messages about their journey and their progress, can also help us think about how a scheduler needs to communicate the milestones in a project.

Every journey has a beginning and an end. Therefore, each project should have a starting and ending milestone. At different points in the journey, the traveler may want to know how close they are to specific landmarks. Each project should report when its landmarks are near or have been passed. To describe the journey to others, the traveler may want to summarize the time needed to travel between specific milestones. Each project should also be able to summarize the plan and actual progress made along the way, without having to recount all the details of the trip. The following paragraphs describe how to define fixed schedule milestones, create and use reporting milestones, and summarize progress with "hammock" activities.

Project Milestones

In the context of construction scheduling, fixed milestones are the most important time-based markers between which a project, or a phase of a project, must start and finish. The construction contract is the primary source of milestones in a construction schedule. Two milestones that will be present in virtually all construction contracts are the contract's Notice to Proceed (NTP) date and the Contract Completion Date.

The first implication of these overall project milestones is that there is one NTP date and one Contract Completion Date on most contracts. This is also consistent with the previous definition of a project as a set of activities that has a start and a finish. The first effective scheduling rule is that there should be a single activity to start the project and single activity to end the project. These activities can be simply called "Start Project" and "Complete Project." Each of these activities will have a zero-day duration to ensure that such nonwork activities do not otherwise impact the CPM calculation. If contract requirements are to be reflected in the schedule, then the NTP and Contract Completion Dates must be somehow assigned to these activities.

The NTP date fixes the date from which all scheduling calculations may proceed. If a project plan is developed on any other day than the NTP date, then the scheduler will have to fix the start of the "Start Project" activity to the NTP. Activities can be said to "start no earlier than" the NTP date. Since the CPM Forward Pass procedure defines the earliest an activity can start, the "Start No Earlier Than" milestone date sets the Early Start Date of the assigned activity. With the Start No Earlier Than milestone, the "Start Project" activity could begin later than NTP, but it cannot begin earlier.

The Contract Completion Date fixes the date by which all work on the project must be completed. The Contract Completion Date must be set to fix the end of the "Complete Project" activity. Activities can be said to "finish no later than" the Contract Completion Date. Since the CPM Backward Pass procedure defines the latest an activity can finish, the "Finish No Later Than" milestone date sets the Late Finish Date of the assigned activity. With the Finish No Later Than milestone, the "Finish Project" activity could finish before the Contract Completion Date, but it cannot end later.

There is little controversy in the application of a Start No Earlier Than milestone for a "Start Project" activity. There is some controversy, however, regarding the use of the Finish

No Later Than milestone for the "Finish Project" activity. This controversy stems from two different lines of argument. First is that the Finish No Later Than will result in negative float when the Early Finish Date of the "Finish Project" activity occurs after the milestone. In this case, the CPM Tutor would be very clear that the purpose of a schedule is to communicate about the project, regardless of the news. If the schedule indicates that the project is behind, by showing negative float, then the project team should determine why and adjust the schedule accordingly. Hiding the fact that a project is behind schedule is a recipe for disaster.

The second controversy, regarding the application of the Finish No Later Than milestone on the "Finish Project" activity, pertains to contractors who plan to finish a job early. Owners are often unwilling to allow contractors to show an early completion since they not want to take the risk for claims if a contractor's planned early completion is delayed. Such owners may even mandate, by contract, that the contractor's schedule show a contract completion date a fixed Finish No Later Than milestone.

Contractors, on the other hand, want to be allowed to finish early and do not want to be constrained to complete the project only on the Contract Completion Date. Such contractors may wish to have the "Finish No Later Than" milestone fixed to their planned contract completion—a date earlier than the Contract Completion Date. It is true that if such schedules are reasonable and the contractor is delayed by the owner, there may be grounds for compensable delays.

The CPM Tutor considers this matter not to be a controversy at all. The typical construction contract in the United States allows contractors to finish the project at any time before the Contract Completion Date. As a result, contractors must schedule their projects to reflect both their plan and the actual prosecution of the work. At the same time, there is a fixed limit to the time permitted on most construction contracts, as expressed in the Contract Completion Date.

Since both positions are correct, it is the recommendation of the CPM Tutor that every project has a "Finish Project" activity with a Finish No Later Than milestone set to the Contract

Completion Date. If the contractor plans a project to complete before the Contract Completion Date, they should also add a zero duration "Planned Finish" activity to the schedule. This Planned Finish must also have a Finish No Later Than Milestone reflecting the contractor's plan.

On some contracts, intermediate, contract-based milestones may also need to be added to the schedule. Such milestones may be included in the project in several ways, depending on the specifics of the contract. If the contract has firm dates for access to specific areas of work, then intermediate "starting" activities should be included in the schedule. The Early Start Dates of such activities should be set to the associated contractual requirement. Similarly, if there are contractual requirements for handover of specific sections of the project, then interim completion activities should be included in the schedule. The Late Finish Dates of these activities should be set to the associated contractual interim completion dates.

In some instances, contracts may also require that a given phase of work begin on a specific calendar date. To include such a rare case, begin by adding a named activity associated with this requirement. Next, the schedule calculation will have to fix both the Early Start Date and the Late Finish Date of an activity.

The inclusion milestone dates in a schedule may have unexpected side effects on schedule calculation. For example, consider the inclusion of a schedule with both a "Planned Finish" and "Finish Project" milestone. Further, let's assume that the initial schedule has a "Planned Finish" date that is 30 days earlier than the "Finish Project" milestone. In this initial schedule, the Total Float of Critical Path leading up to the "Planned Finish" activity should be zero. The Critical Path extends to the last activity, "Finish Project," has a 30-day Total Float.

Now consider that the project is under way and the contractor's initially optimistic schedule has slipped by 15 days. This means that the contractor will complete 15 days past their "Planned Finish" date. Thus, the Total Float of the Critical Path leading up to the "Planned Finish" activity would now have a value of −15. Since the "Finish Project" activity was originally scheduled to be 30 days later than the "Planned Finish," the

Total Float on the "Finish Project" activity would change from 30 to 15 days.

While having a Critical Path whose activities have different Total Float values may seem incorrect, it is not. The definition of the Critical Path as, "the set of activities with the least Total Float," does not require Total Float values to be equal! Anytime there are milestone dates such differences are the norm, not the exception. Therefore when the scheduler comes across float values that change, a "plugged date" or milestone should be expected.

Progress Milestones

The use of milestones that fix an activity's Early Start Date and/or Late Finish Date have a significant effect on CPM calculation. To minimize the team's effort to recognize and understand the impact of such fixed milestone dates, the CPM Tutor recommends that milestones only be added to contractually specified activities. There are, however, specific stages of the project where the team will need information about key progress (versus contractual) milestones. This can be accomplished with a "Progress Milestone."

A Progress Milestone is simply an activity with zero duration inserted into a schedule at the start or end of a given series of activities. With a zero duration, the Progress Milestone will not affect the CPM calculation. The start and finish of the activity will move with the schedule. The CPM Tutor recommends that if such activities are needed, they should be named with the words "Start" or "Finish" prepended to the name of the work to be reported.

Hammock Activities

During the project, many stakeholders will want to know the general progress of the project without getting into the details. For schedules with thousands of activities, it is often very difficult to summarize the progress of the work without creating and reporting on many Progress Milestone activities. Another tool in the schedulers' communication arsenal is the Hammock Activity.

As the name implies, the Hammock Activity swings between the Early Start Date of the preceding activity and the Late Finish Date of the following activity. The Hammock Activity has no duration of its own, but takes its duration from the difference between its preceding and following activities. Hammock Activities are extremely useful in reporting a variety of different types of progress. These would include, but not be limited to, reporting the starting and ending of work on each floor of a high-rise building, or a subcontractor's window for coming on and off the job.

Activities

The ancients knew about roads and also how to get there. Lao Tzu's famous quotation is, "A journey of a thousand miles begins with a single step." Once the project gets going, the second, third, and remaining steps are also required. To the construction scheduler, the steps needed to complete the project are the activities that show progress between milestones. In Chapters 3 and 4, it was assumed that a set of activities for a project has been defined. In the following paragraphs, the CPM Tutor will provide best-practices about how to effectively model construction activities.

Activity Coding

As the rules for developing good schedules are developed, starting in the paragraphs below, there is a feature of CPM scheduling programs that should first be introduced. That feature is called "Activity Coding." Activity Coding is of critical importance when developing large schedules. More generally, Activity Coding is a set of classifications that allows the scheduler and stakeholder to select, and review or update, specific activity subsets.

Greg Ceton at the United States Construction Specification Institute provided a very good example of the importance of such classifications in a 2010 presentation attended by the CPM Tutor. Consider, for a moment a giant box of different types of balls. If one wanted to only look at specific kinds of balls, how would he/she explain to a helper which ones to retrieve? There could be several different ways to make such a request. For example, someone interested in recreation might select, "all balls used for baseball" or "balls used for basketball." A supplier of materials might select, "all balls made of leather" or "all balls with a liquid center." A shipper may only be interested in balls above a specific weight and volume, inflated, and deflated. Another person may be interested in balls of different shapes, or ones made of a single material, or ones that were produced using a specific stitching machine. For a person answering such queries, any and all of these requests could be valid.

Just like the boxes of balls, a large set of activities can be described using different types of Activity Codes. Activity Codes allow the scheduler, the project partners, and stakeholders to select a subset of activities pertaining to specific items of interest even in schedules of thousands of activities. Activity Coding should be assigned to activities when the activity is first created and travel with the activity as the project progresses. Many owners have specific requirements for Activity Codes and some of these may be directly identified by contract. It is a CPM Tutor best practice to develop and use standard Activity Coding on all projects, even smaller projects.

Some readers may be familiar with the term "Work Breakdown Structure." Activity Coding is presented in this book as a set of discrete Activity Codes instead of a large (and often confusing and inconsistent) ontology. A Work Breakdown Structure, in the view of the CPM Tutor, is simply the compiled set of Activity Codes used on one specific project.

Means and Method

The first step in defining what activities are needed is to understand how the project is to be constructed. The plans and specifications themselves mandate those specific to activities be included, or not included, in a schedule. For example, if an exterior wall is to be built of brick and masonry block, the activities needed differ from a project for which the exterior closure is provided by a glass curtain wall. Once the type of construction is identified, contractor and subcontractor expertise with specific construction techniques and local methods will govern the set of activities to be included in a schedule. In relatively low-wage countries, for example, activities dependent upon human labor will be added to the schedule in lieu of activities that substitute machinery for labor.

The decision about means and methods is often implicit. For example, some contractors and subcontractors specialize in one set of means and methods to capture specific types of business. The scheduler should have a conversation about the means and methods to be used by the contractor and subcontractors. If not discussed, the scheduler may be able to determine this information from bid estimates.

There is no one perfect formula to define construction schedule activities. Given a type of construction, and a defined set of means and methods, the activities needed for a project can be defined many different ways. The CPM Tutor defines an activity as a set of work accomplished by a single type of crew, in a given work area, with a defined set of tools and materials.

Activity Coding should be used by the project team to help define, select, and report on activities to be accomplished by a given crew (and or subcontractor), in a given work area, and of a given type of work. Crew type is typically defined in the "Responsibility Code." Work area is defined in a "Work Area Code." The type of work being accomplished is described in a "Work Type Code." At the start of the project, the allowed values for each code category should be specified by the scheduler. In some cases, the coding values may even be mandated by construction contract. The simple application of these codes will provide big payoffs when evaluating the schedule for productivity. Productivity issues are specifically addressed in Chapter 7 under the Unlimited Resource Leveling heading.

Work Areas

The most effective means to define activities is to look at the flow of specific types of activities (and their crews) through the work areas in a project. In most projects, these work areas are evident. Buildings, for example, may be described by work on different floors, wings, or zones. Activities on multibuilding projects can be defined by building and then by floors, wings, or zones. Horizontal projects also have distinct sections, or reaches, that can be used to partition similar work.

Activities identified by Work Area Code can be selected from all activities and included on specific reports. For example, if the project team wanted to find only those activities pertaining to the third-floor, west-wing, of building 2, the Work Area Code could select only those activities.

Hammock activities are often added to the first and last activities in given work areas. For example, in a high-rise building, it may be important to complete the first several floors of the building for retail and office rentals as quickly as possible, while work continues to finish upper floors. Adding a Hammock to the start and end of early occupancy floors would be very helpful for project stakeholders.

Duration

The duration of a given activity should be based on the amount of work to be accomplished and the expected productivity rate of the crew. In most types of schedules, the productivity rate of a given type of work may be assumed to be the same regardless of the crew. Because of this, the productivity rates used for the construction cost estimates may be a helpful resource to determine an activity's duration.

Of critical concern when developing an activity duration is to consider how to report progress on that activity. Progress is typically reported on a monthly basis. To simplify the discussion with the owner about assessing progress, it is critical that the schedule be a clear communication tool and not a source of contention. Agreements by all parties to monthly schedule updates allow the schedule to be an effective tool for project payment. To ensure that communication about progress is clear, the duration of a construction activity should never exceed the monthly reporting period. Most activities with such durations will begin and end during a single month. Those that do not complete during the first month can be expected to complete the next month. Even if there is some disagreement on the progress of in-progress activities during the current month, completing the activity the next pay period will quickly resolve any disputes.

Calendars

Real projects do not begin on day zero and complete on workday 13, as was the case with the examples provided in

Chapters 3 and 4. Real schedules begin and end on specific calendar days. Crews work only on specific workdays for specific hours, with specific paid holidays. Crews may, in fact, work on different schedules with different workdays and hours. Keeping such information straight is one of the important capabilities of computerized CPM algorithms. The paragraphs in this section describe how these algorithms use calendars.

In Western countries, construction crews typically work 8-hour shifts, Monday through Friday. To evaluate the impact of such a workday on CPM calculations with calendars, consider an activity with a 1-day duration. According to the CPM calculation provided in previous chapters, if the activity was to have an Early Start Date of day 10, its Early Finish Date would be day 11. Now consider if calendars are added to the schedule. If the activity has an Early Start Date of Apr 15, 2014, and there is an 8-hour workday, then the activity also has Early Finish Date of Apr 15, 2014. Here, we see that the Forward Pass calculation is not Early Start Date plus activity duration. The calculation is Early Start Date, plus duration, minus 1 workday.

A similar change must be made to the Backward Pass calculation when using calendars. If, for example, there was an activity with a Late Finish Date of Apr 15, 2014, its Late Start Date would not be the Late Finish Date minus the duration of the activity. The Backward Pass would set the Late Start Date to be the Late Finish Date, minus the activity duration, plus 1 workday.

When propagating Early Finish Dates forward to following activities, the noncalendar calculation would set the Early Finish Date as the Early Start Date of the following activity. With calendars this must be adjusted. If an activity has an Early Finish Date of Apr 15, 2014 and one follower, then the Early Start Date of the follower would be Apr 16, 2014. That is the Early Finish Date of the prior plus 1 workday. Similarly, the Backward Pass with calendars is also adjusted by a minus 1 workday.

Most crews work on a 5-day workweek. Again, computerized CPM algorithms take care of this bookkeeping. Assuming a Western calendar, with Saturday and Sunday as non-workdays, if an activity has an Early Finish Date that falls on a Friday, then the next available start is not the Early Finish Date plus 1 calendar day. The Early Start Date of the following activity is the Early Finish Date plus 1 workday, moving the start of the activity to the following Monday.

A specialized type of non-workday is the holiday. As far as computerized CPM algorithms are concerned, holidays are simply another non-workday. The calculation is the same as that described previously. The inclusion of holidays in calendars is a small technical detail but consider its importance when communicating to members of the construction team. If the plumbing crew mistakenly found out that they were to start the second floor fixture installation on Friday, Jul 4, 2014 (a United States Federal Holiday), both the scheduler and the schedule would be discredited.

A 5-day workweek is not the only type of workweek that exists on most construction projects. For example, many administrative activities have calendar day durations. The amount of time allowed for an owner's representative to review a shop drawing, for example, may be specified in the construction contract's general provisions as being 45 calendar days. Other types of nonwork activities, for example concrete curing, will also take place based on 7-day workweeks.

Total Float calculation is also affected by the calendar of the activity. The difference in Late Finish Date and Early Finish Date to calculate Total Float only makes sense if the calculation is done when taking into account the activity's workweek.

Now consider the increased complexity in fully understanding schedules which have both 5- and 7-day workweeks, and holidays. In some cases, the Critical Path could run through projects which have different calendars assigned to them. These activities may all have the same amount of Total Float; however, some of the activities could be accelerated if needed by working weekends. Other activities on a 7-day workweek may not be able to be accelerated.

The CPM Tutor recommends that most projects use one 5-day workweek calendar, and necessary holidays, as their primary calendar. The use of an additional 7-day workweek calendar should be restricted to only those activities that specifically require it. Administrative and evaporative activities are examples of activities that require a 7-day workweek. The scheduler must keep the list of activities, and the explicit justification, with alternate schedules in their minds when developing and reviewing the schedule. If the scheduler does not have simple rules for the application of alternate calendars, then a written memo accompanying the schedule listing these activities should be provided with each schedule update to inform the project team.

Activities and Operations

Within large, complex projects it is often difficult to know how to partition different types of work into manageable activities. The scheduler's dilemma is how to break work into small enough parts to be effectively managed without making the burden of evaluating progress too great. As a result, activities are typically defined as units of construction accomplished by workers of the same type, using the same tools and materials, and working in the same physical work area. Activities, like projects, have a beginning and end.

Separating activities by crew type ensures that upstream or downstream problems, or changes to subcontracting scopes, will be included in the schedule without a major revision. If a given crew can perform multiple types of work, when they change between one work type and another, there may be problems with getting materials or tools to the workface. As a result, if a crew changes the type of work it performs, the work should be identified as a separate activity. Finally, crews complete work through a series of work areas. Having multiple crews competing for a single work area, or having crews do work in a given area out of sequence, is a recipe for tear-out and rework.

As with any set of rules-of-thumb there are exceptions. For example, some practicing schedulers often combine different types of crews into a single activity. The CPM Tutor recommends that these exceptions be explicit decisions to deviate from the clear definition of an activity. While the recommended schedule will have more activities than schedules that combine multiple crews into a single activity, the techniques in this book will make such larger schedules very manageable.

It is also important to note that activities are composed of multiple individual construction operations. Construction operations are of a much smaller scale than construction activities. Operations are discrete motions or actions taken by a single crew at a scale of minutes or hours. Unlike construction activities, operations are cyclic. Cycles are patterns of work repeated over and over by individual crews. These cycles deliver materials to a workface and install that material using labor and equipment. Evaluating such cycles is vital for productivity improvement, but because the scale of the analysis is in minutes or hours and the nature of the work is cyclic, CPM is not the most representative model of individual, detailed construction operations.

Sequence

With a list of activities and durations, the next set of information to be added to a schedule is the sequence. The scheduler will find that there is an interplay between the list of activities, their crews, work areas, and sequence that will require further adjustment. In the following paragraphs techniques to ensure that a schedule effectively reflects the logic between activities are discussed.

Phasing

When defining activities, one of the first considerations is that of the fixed schedule start and end dates in the contract. Additional, intermediate milestones may also be identified if specific phasing requirements are included in the contract. When developing the sequence of the schedule, these intermediate milestones need to be considered. The scheduler should ensure that there is one beginning and one ending activity for each required phase and that the logic of activities in that phase is consistent with the contractor's plan to complete the phase. To the extent possible, phases of work should be completed as independently as possible. This will allow the schedule to be clear about progress of that specifically required phase. Delays and changes to activities in a given phase of work are easier to understand and explain if activities in that phase can be developed as a mini schedule within the overall schedule. Some schedulers refer to such mini schedules as "frag-nets" or fragments of networks.

Activities identified by Phase Code may be selected from all activities and included on specific reports. For example, if the project team wanted to find only those activities pertaining to "Phase 1," the Phase Code would be able to select only those activities.

Crew Chase

In Chapter 2, A History of Project Planning, it was noted that Bar Charts were based on attempts to measure productivity in factories. In factories, a work piece moves on a conveying system through a variety of stations where the production resources (workers and machines) attach premanufactured components or assemblies to the in-progress piece. At the end of the line, after all components have been attached and the product has been tested, the product is ready for sale.

Construction processes could also be considered to be a type of manufacturing. The construction process, however, is significantly different from that of the assembly line. This difference results from the size of the product being built. The size of the product requires that workers to come to the product. This is the opposite of factory production where the partially completed product comes to the workers. On a construction site, workers move across the partially completed product, in a specific order, according to the technological constraints of the production process, their production rate, and the availability of material and equipment.

An experienced contractor and scheduler can visualize this flow of workers across the project. From this point of view, schedule logic simply reflects the flow of crews across the project, one following another, through each needed work area. This flow of crews is called the "Crew Chase." In the view of the CPM Tutor, the Crew Chase is a fundamental concept in the formation of an effective schedule.

As the scheduler considers the crews moving through a given work area, the set of activities becomes clear. In the example of the construction of one floor of a mid-rise Cast-in-Place Concrete building, the set of activities might be as follows: formwork, place concrete, exterior closure, stud wall, rough-in electrical, rough-in plumbing, HVAC, finish carpentry, finish electrical, finish plumbing, paint, and carpet.

With this set of activities, the scheduler can envision crews moving not only from one floor but through the entire building across all floors. The formwork moves vertically up the building followed by each of the other types of activities, in order. To ensure that work can be planned as soon as allowed, but no sooner than possible, the sequence between activities has to take into account several different types of constraints.

The sequence of the schedule must enforce three constraints governing how soon the crew can move from one work area to the next. The first is constraint is physical. The second floor cannot be built until the first floor is built, and the structure is strong enough to support the weight of workers and equipment. The second constraint is capacity. The space has to be clear so that the new work can begin. The third constraint is preferred order of efficient operations. Carpet is completed after painting to ensure that a "Clean Carpet" activity is not required to remove spilled paint.

Practically speaking, these constraints are translated into the CPM sequence in three ways. First, within a given work area, the preferred order of efficient operations is determined and scheduled. Next, the set of activities is repeated through each similar work area. Assuming one crew or set of crews will perform these activities, the activities accomplished by each crew are also linked by activity logic. Finally, any technological constraints, such as requirements for concrete curing or positioning equipment, also have to be added to help plan, and alleviate, if possible, any impacts of such technology.

The scheduler can employ Activity Coding to summarize and simplify the workflow and associated constraints for the project team. The simplest type of Activity Coding is in the numbering of the PDM activities themselves. PDM activity numbers are, except for the requirement for increasing activity numbering, arbitrary. This means that the PDM activity number itself can be used to help organize the schedule. In computerized PDM programs, the activity number may also be called the "Activity ID."

Consider, for example, the use of an Activity ID composed of six characters. The first two numbers can be used to signify the work areas. The starting numbers of "00" can be used for site work, while the ending numbers of "99" can be used for any closeout activities. Other numbers could reflect, in the example of a mid-rise building, the floor number. Thus, "02" would be the second floor. For a 10-story mid-rise building, an Activity ID of "11" would represent the roof.

The next two digits could be set to be the type of work being performed. These digits would increase, for example, from "0105," for the first-floor formwork, to "0165" for the first-floor carpet. On the third floor of the same building, an Activity ID starting with "0305," can be seen by inspection to be the third-floor formwork. Activity ID "0465" is the fourth-floor carpet.

The final two digits might be used to indicate the subcontractor or crew performing the work. Since one crew or subcontractor might preform multiple types of work on a job, identifying activities by Responsibility Code will be helpful to evaluate progress of specific aspects of the project. Typically, the contractor creating the schedule will define Responsibility Code by the prime contractor crew type names and the name of the subcontracted work package. This is because the prime contractor creating the schedule may not have specific information about, or responsibility for, subcontracted crews.

As the scheduler begins to develop the sequence of activities, they further refine the specifics of some activities and combine others. That result should be simple and clear allowing users to understand the schedule without needing a translator. This can be accomplished using the Activity IDs, Activity Codes, and Hammocks. The Activity ID approach will be most useful when specific activities need to be found and reviewed on activity reports. The Activity Coding approach will be most useful when performing schedule analysis. Hammock activities will also be needed to summarize the schedule for project stakeholders.

The Activity Codes to describe the Crew Chase are the Work Area Code and the Responsibility Code. The Work Area Code is a value assigned to each activity that identifies the areas of work through which crews follow one another. The Responsibility Code is a value assigned to an activity that identifies the type of crew, or subcontractor, responsible for the given activity. Since the same workers could do several different types of work an additional Activity Code, Work Type, helps to identify the specific contractor or subcontractor crews as they move through the project.

Hammock Activities are activities which have no duration of their own. These activities "swing" between the Early Dates of the preceding activity and the Late Dates of the following activity. With Hammocks, the scheduler can produce a simple, one-page summary report of a schedule comprised of several thousand activities. Once such summary report, linking the first and last activities in a given floor of a building could reduce the schedule of a 10-story building to a report, or even a Bar Chart, containing as few as 13 activities. There would be one Hammock Activity for mobilization and site activities, one Hammock Activity for each floor, one Hammock Activity for the roof and associated equipment, and one Hammock Activity for clean-up and demobilization activities.

Templates

Walter Boyce in his 1991 article "Designing for Constructability" talked about the need of the builder and the contractor to design constructible buildings by keeping the design straight having consistent spacing and sizing. The resulting workflow in buildings with such designs is that the work to be accomplished is predictable and repetitive. Given a typical building, the schedule will also be predictable and repetitive. The schedule can reflect the crews as they move from one work area to another.

There are standard patterns of work in construction. These patterns are not perfectly applicable to all buildings of all types; however, standard patterns are often sufficiently representative to be useful to the scheduler. A good pattern will express both the physical, means and methods, and operational constraints among its activities. Taken together, standard patterns can be used as templates for schedules, or parts of schedules, from which a more detailed conversation about the specifics of a project can begin. Since getting started with any new project is often the hardest part, having a place to begin may be helpful for those new to scheduling.

One important set of patterns that could be applied in many situations is developed from the construction of types of structural systems. The patterns provided give one possible list of "typical" activities for a structural steel building, a precast concrete building, and a Cast-in-Place Concrete building.

Table 5.1 provides the list of activities for a generic structural steel framed building. The first column in the table captures the work area where the activity applies. The second column provides an example Activity ID that reflects where the activity would go in the context of other activities in that work area. The third column provides the Activity Name. In this generic template, there are two general work areas—"General Conditions" and "Per Floor." The Work Area Code for the first three activities in the template could be set to "General Conditions" as these activities take place prior to any construction, but do impact the start of the construction. The remaining activities in the list occur on each floor as the building is constructed. The example Activity IDs provided in the second column reflect both the Work Area Code and provide a specific identifier for the activity. The first two digits of the three General Conditions activities are set to "00" to reflect that these activities precede work on the first floor. The second of the four digits in the example Activity IDs provided are an arbitrary set of ID numbers. The

Work Area (Example)	Activity ID (Example)	Activity Name
General Conditions	0030	Submit Steel Shop Drawings
General Conditions	0031	Approve Steel Shop Drawings
General Conditions	0032	Procure Steel
Per Floor	0205	Deliver Steel
Per Floor	0210	Erect Steel
Per Floor	0215	Install Metal Deck
Per Floor	0220	Install Deck Slab
Per Floor	0250	Fireproof Steel

Table 5.1 Steel Frame Template

only suggestion provided for these Activity IDs is that they be sequential so that the scheduler can easily remember how to find the General Conditions activities. The scheduler may even wish to add a Hammock activity if the procurement of the steel is on the schedule's Critical Path.

The "Per Floor" activities in Table 5.1 follow one another through each floor as the building rises. The only exception to this rule is the activity "Deliver Steel." Given that the cost of the structural steel may be a relatively large portion of the job, it may be possible to have the delivery of the steel for multiple floors occur early in the project. In other cases, projects may not have the luxury of a large lay-down yard and require the delivery of steel on a just-in-time basis. Regardless of which approach is needed, the sequence of "Deliver Steel" can be adapted to the project.

The next three activities, the ones that that actually build the structure, "Erect Steel," "Install Metal Deck," and "Install Deck Slab," are typically accomplished in order, but may be done by different crews. Since these are accomplished by different crews, Erect Steel can be moving forward on floors above where the Install Deck Slab activity is working. These crews may share elevators and/or cranes so may be constrained in a number of different ways depending upon the means and methods used for each project.

The last activity in the list "Fireproof Steel" does not typically occur directly after "Install Deck Slab." This is because the fireproofing will often only be applied following the completion of some, but not all, other activities. Early application of the relatively delicate and continuous fireproofing will require reapplication.

The example Activity IDs provided give a general idea of the sequence of the Steel Frame Template activities. The example Activity IDs of the last five activities in Table 5.1 comprise the work to be accomplished on each floor. The first two digits of these activity IDs show the example coding, "02", for activities on the second floor. The first four of these activities will typically be accomplished directly in order since no other work on that floor can be done without there actually being a floor. The Activity IDs of these first four activities are in close sequential order (with gaps allowing any possible future changes). The Activity ID of "Fireproof Steel" is not closely sequential to reflect the fact that fireproofing must take place later in the sequence, after rough-in utilities, but before ceiling and possibly before wall closure.

The next template to consider is for a Cast-in-Place Concrete facility as shown in Table 5.2. As in the case of a structural steel building, there are activities that may, or may not, need to be included in the schedule pertaining to General Conditions. The decision to include these, or not, will be completely based on the job and local conditions. Additional discussion about when to include, or not include, such activities can be found later in this chapter.

As with the Steel Structure Template, the Cast-in-Place Concrete Template will need to be adjusted, depending on the specifics of the methods used and project requirements. For example, the activity "Deliver Rebar" may occur for each floor or in batches larger than one floor. The impact of bucketed- or pumped-concreting will also need to be addressed. A requirement that is more important

Work Area (Example)	Activity ID (Example)	Activity Name
General Conditions	0030	Submit Mix Design & Certification
General Conditions	0031	Submit Rebar Certification
General Conditions	0032	Submit Concrete Placement Plan
General Conditions	0033	Approve Concrete Submittal
General Conditions	0034	Procure Concrete Materials
General Conditions	0035	Deliver Concrete Forms
Per Floor	0205	Install Concrete Forms
Per Floor	0210	Deliver Rebar
Per Floor	0215	Install Rebar
Per Floor	0220	Place Concrete
Per Floor	0225	Finish Concrete
Per Floor	0230	Cure Concrete
Per Floor	0235	Strip Concrete Forms
Per Floor	0285	Move Concrete Forms

TABLE 5.2 Cast-in-Place Concrete Template

to the Cast-in-Place Concrete Template compared with the Steel Structure Template relates to the task "Cure Concrete."

In the Cast-in-Place Concrete Template, the time to cure concrete needs to be included since work above a floor must wait, usually 2 weeks between floors for the concrete to cure. At the conclusion of the work on a floor, the movement of forms between floors may also have, or not have, a constraining impact on the floor above. As with any example template, the application of the generic template should take into account all project-specific constraints.

The next template shown in Table 5.3, for Precast Concrete, can be used to develop activities within or outside the facility. The application of the Precast Concrete Template, like any other template, will be project specific. This template, like the previous ones, has activities that pertain to both the General Conditions and the individual Work Area.

Once the structure has been completed, crews can, one-by-one, move through these work areas to fill in the walls, windows, doors, services, and finishes to complete each floor. Table 5.4 shows a generic template for the placement of walls on a given floor. The most important aspect of this template is that the work needed to install partitions requires detailed collaboration among trades. For example, metal studs need to be framed to support both the drywall but also to support electrical

Work Area (Example)	Activity ID (Example)	Activity Name
General Conditions	0040	Submit Precast Drawings
General Conditions	0041	Approve Precast Drawings
General Conditions	0042	Fabricate Precast
General Conditions	0043	Deliver Precast
General Conditions	0044	Deliver Crane
Per Work Area	0250	Install Precast Panels
Per Work Area	0255	Install Tie Strips
Per Work Area	0260	Install Caulking
Per Work Area	0265	Seal Precast

TABLE 5.3 Precast Concrete Template

Work Area (Example)	Activity ID (Example)	Activity Name
General Conditions	0052	Submit Drywall Partition System
General Conditions	0053	Approve Drywall System
General Conditions	0054	Procure Drywall System
General Conditions	0055	Deliver Drywall System
Per Work Area	0240	Install Metal Studs
Per Work Area	0245	Rough Plumbing
Per Work Area	0250	Rough Electrical
Per Work Area	0255	Rough Communications
Per Work Area	0260	Install Drywall
Per Work Area	0265	Tape and Finish Drywall

Table 5.4 Drywall Partition Template

outlets, pipe and duct penetrations, the installation of dampers, and access panels. Some contractors will place the studs and one side of a drywall wall, others will simply place the studs first. Separating activities into work accomplished by a single crew in a given work area will help the contractor reduce tear-out and rework and reduce crews crowding into the same work area.

The generic installation of any discrete product or piece of equipment can be developed into a template. Table 5.5 provides a generic set of activities relevant for almost any large product. Such a generic product template could even be used to further categorize the activities in the schedule through the use of Work Type Code. This Work Type Code, in Table 5.5, reflected in a new six-digit style of Activity ID.

Work Area (Example)	Activity ID (Example)	Activity Name
General Conditions	0001XX	Submit
General Conditions	0002XX	Approve
General Conditions	0003XX	Procure
General Conditions	0004XX	Deliver
Per Work Area	0201XX	Install
General Conditions	0202XX	Test

Table 5.5 General Activity Template

Given that virtually all products will proceed through these steps, the Activity ID has been expanded to six digits. As before, the first two digits represent the Work Area Code. The value of "00" represents General Conditions. The value of "02" is representative of the second floor. The next two digits could be used to represent the Work Type Code. In Table 5.5, the "Approve" activity for any kind of product would be identified as a "02." The approval of concrete, for example, could be set to "000225" while the approval of rebar could be assigned to a value of "000230." Since the last two digits of the activities shown in Table 5.5 would be used for any product, or material, these values are shown in Table 5.5 with the wild-cards "XX." The application of the above nomenclature is by no means standard; however, any technique the scheduler can use to organize large sets of activities into patterns that can be readily developed and understood will be important. Experienced readers of this book will no doubt have their own favorite methods of numbering and coding activities.

When developing a sequence of activities, experienced schedulers use some important rules-of-thumb. These rules-of-thumb, like the templates above, must be considered within the context of each project. These rules-of-thumb are outlined and the justification for each are provided below.

1. Framing and the interior skin of mechanical rooms should be scheduled as early as possible.

   The rationale for this requirement is to provide access for mechanical installation crews whose work is tightly constrained within small mechanical rooms. Finishing the inner side of the walls of mechanical rooms also allows ductwork, equipment, and conduit to be placed on those walls without the finishers and painters needing to come back to the room. The outer side of the wall should be left open to allow framing for sleeves and dampers as needed once the ductwork has been placed during rough-in.

2. Exterior closure should incorporate separate activities for building "skin" and window installation.

Depending on the local weather conditions, it may or may not be relevant to manage the exterior closure milestone of a project. If weather that may impact construction is possible, then precautions may be taken by covering exterior wall openings if windows have not been installed. In addition, the early installation of windows may result in excess breakage.

3. In some cases, sealing and caulking the exterior after window installation should be included.

   The development of a tight building demands that extreme care be taken when finally fitting up and closing the building. As a result, a separate activity may be warranted to be followed by any necessary testing.

4. In high- and mid-rise buildings, the building skin of the first and second floors should, generally, be scheduled last.

   The areas where workers travel and move material and equipment will be more prone to wall scrapes and window breakage than areas of the project not used for staging construction operations. Since the lower floors are often used for these staging areas, they should, in absence of other constraints, such as early leases on street-level retail spaces, be scheduled last. If access to the facility through lower floors is constrained, then other accommodations will need to be made to ease worker access.

5. In high- and mid-rise buildings, or horizontal projects, installation activities should, generally, be defined by templates repeated on every similar work area, such as across building floors or road mile markers.

   A "constructable" schedule will be easier to explain and use. As will be shown in Chapter 7, such a schedule can also support a variety of very sophisticated types of analysis.

Cost-Loaded Schedules

Many contracts require that construction activities be "Cost Loaded." From the point of view of the owner, such "costs" are more correctly described as a proportion of the fully burdened construction contract amount allocated to each construction activity. Cost Loaded, for the owner, is not the contractor or subcontractor cost at all, but the cost of that activity to the owner. Although the CPM Tutor would prefer the terminology "Value Loaded" schedules, the term Cost Loaded is discussed because of its widespread use in the industry today.

The first rule of Cost-Loaded schedules is that the sum of these costs for all activities must equal, to the penny, the total construction contract amount. Furthermore, many contracts require the contractor to report earnings against a contract-specific line item or "Bid Item Code." The Bid Item Code is always required by owners when projects are funded by using different owner cost accounts. The sum of the values for each Bid Item Code will be required to match the total contract amount for that contract line item.

Since it would be too difficult to include and allocate different percentages to activities with multiple Bid Item Codes, the CPM Tutor strongly recommends that every activity be assigned only one Bid Item Code. Activities with a Cost-Loaded value of zero which would be appropriate to any nonwork activity should have a Bid Item Code set equal to "n/a." Ensuring that there is an explicit Bid Item Code for every activity will help the scheduler check that items were not left off the list by mistake.

As a general rules-of-thumb, the CPM Tutor recommends that there are two ways to develop Cost-Load schedules. The first way is to assign a consistent cost-per-day Cost-Loading to every activity that is performing the same work. Such a consistent cost-per-day Cost-Loading will be very easy to explain to all project partners, and typically results in a schedule of payments that is agreed to prior to starting work. Creating a schedule using templates greatly simplifies this process since the cost allocation rate will be the same across all similar template activities—even if the specific quantity of work for each activity differs slightly between work areas.

The second allocation of costs in Cost-Loaded schedules relate to activities that include the installation of large equipment. In these activities, the value of the activity to the owner is primarily achieved when the equipment is delivered to the jobsite. The installation of that equipment, while critical to the operation of the facility, is fiscally less important than the actual delivery of that gear.

In such cases, the CPM Tutor recommends that a specific "Deliver Equipment" activity be added to the schedule. This type of activity assists the contractor to obtain valid payment for the equipment as quickly as possible.

After the scheduler has, in conjunction with the contractor, allocated costs to each relevant work and equipment delivery activity, a summary report should be developed. Many computerized CPM scheduling programs will also produce a graphical report of planned earnings. As a double check that planned earnings are reasonable, the graphic report should take the familiar S-curve shape first introduced in Fig. 2.11.

In addition to simplifying the owner's and possibly the contractor's cost accounting, the Bid Item Code has one additional, and critical, function. That function is to specifically identify activities that are added to schedules to justify cost and time extensions for Requested or Proposed Changes. If activities that need to be added to a schedule have a Bid Item Code, such as "RFP 125" or "Change Order 1," then all project partners may be able to more clearly assess the impact of such proposed or agreed upon changes to the schedule's duration. The inclusion of the Bid Item Code will also help the owner's representative make the case for the owner to provide additional funding if such is needed to cover the cost of a proposed change. More will be covered on the topic of claims and changes in Chapter 6.

Some contractors may take issue with scheduling at the level of detail described by the CPM Tutor. Such contractors may not want the owner's representative to closely review the contractor's plans and progress. To evaluate this position, consider first that an appropriately Cost-Loaded schedule gets the contractor and owner to agree to a payment schedule. This should result in less controversy when determining monthly billings. Also consider that helping the owner's representative obtain the funding to more quickly execute change orders results in fewer claims after the project is over. Given these considerations, it is the opinion of the CPM Tutor that an effective and transparent schedule results in the contractor doing themselves the favor, not the other way around.

Known Unknowns

In the 1980s, Dr. John Melin, a professor from the University of Illinois, used the phrase "known unknowns" to help students understand potential problems encountered at the construction site. Each of these potential problems is common enough to be well known. The unknown factor of these problems is simply when they would occur on any given project. Good project planners have to be a bit paranoid—they wonder what could go wrong with their projects and then they add activities to the schedule to help them anticipate and resolve issues. The cost of this effort is simply the addition of a few extra activities in a construction schedule.

The management guru Peter Drucker's famous quotation is apropos: "What gets measured gets managed." The inclusion of activities that have the potential to cause delays, or inefficiencies, will alert the project team to problems if such errors arise on the current project. The following paragraphs identify some of these known unknowns.

Equipment

One critical category of a constraining activity is major construction equipment. For some types of construction equipment, such as cranes, the arrival of the crane, or moving the crane from one side of the building to another, will put all other work on hold. Any equipment used to convey labor, materials, or equipment through the project, either vertically or horizontally, should also be closely monitored. The scheduler should include appropriate activities to make sure that all temporary construction equipment is installed and operating correctly.

Administrative Activities

In many instances, actions outside the control of the prime contractor can impact the project. One category of activity is that of administration. The author Annie Dillard remarked that "A schedule defends from chaos and whim." This is certainly the case when trying to anticipate problems associated with administrative activities under the control of the tenants, owner, or owner's representative.

Tenants may delay the selection of building finishes such as carpet and wall colors. The owner may restrict access to specific parts of a facility for some period of time. The owner's representative, or designer, may have an employee's vacation time schedule at the same time the review or approval of a key submittal is required.

The scheduler should, on all projects, determine the past history of administrative delays on previous similar projects. Those items which were the cause of previous delays should be included in the schedule. The intention is not to include every administrative approval, shop drawing review, or catalog cut certification in the schedule; the intention is to include only those which have a high probability and a direct potential impact on activities on or near the Critical Path.

To include such activities in the project, and allow summaries of these activities to be readily available to all parties' Activity Coding is, again, useful. The Responsibility Code need not be restricted to the identification of crew types and subcontractors. The Responsibility Code should also identify the owner or the owner's representative as the party responsible for administrative activities. With a Responsibility Code set to "Owner," the scheduler will be able to generate a report of all activities to be accomplished by the owner. This will help track if the owner is encountering any delays in completing these impactful activities as the work progresses.

The Category of Work code can also help to identify activities. The Category of Work code, according to the U.S. Army Corps of Engineers' regulation ER-1-1-11, is used to identify the general category of work being performed. In the case of administrative activities, such as approvals, the Category of Work code value could be identified as "Admin." Given the list of activities for which the Category of Work is "Admin," the scheduler would be able to determine if additional office-based effort needs to be directed to the owner, subcontractor, or even in-house so that the Critical Path is not unexpectedly delayed.

Procurement Activities

After the approval of a major piece of equipment, it must be ordered and delivered. The delivery of long-lead time equipment may even be on the Critical Path. A common example of such a type of equipment is elevators. Once approved, elevators may require a year (or longer) to be manufactured and delivered. Elevators are typically purchased and installed by the contractor so any delay in the delivery of such equipment will be on the contractor.

Other types of equipment, such as medical diagnostic machines and emergency backup generators, may be purchased directly by the owner. This is because large owners are often able to expedite these purchases in ways that contractors cannot. Unfortunately, for the contractor, the owner who fails to expedite the delivery of such critical equipment may delay the project.

To distinguish between who purchases and who installs such equipment, several acronyms are typically used. The first one is "CFCI." This stands for Contractor-Furnished-Contractor-Installed. This is typically the case with most products, and thus the least likely to be of interest as a separate category. The next one is "OFCI." This stands for Owner-Furnished-Contractor-Installed. This category is the most problematic since the owners' actions may impact the Critical Path. The final category is "OFOI." This stands for Owner-Furnished-Owner-Installed. The equipment furnished and installed by the owner often includes furniture and telecommunication devices installed after beneficial occupancy but before tenants move into the facility.

The OFOI category is also problematic if the owner starts these contracts without proper coordination with the contractor's schedule. For example, the owner could give Notice to Proceed on their contracts while the construction is still taking place. In the worst case, this may result in the furniture installer trying to move furniture into a room that has not yet been painted. Another example could result if the owner's representative does not coordinate the placement of conduit, causing the contractor not to install proper chases or sleeves through walls. In the worst case, this may compromise the fire protection system since the owner's installer may not know which walls are fire rated, and which walls are not.

As with administrative activities, the Responsibility Code and Category of Work may be helpful. The Category of Work code can identify these activities as "Procure." Reports on this category would then give a snapshot of the current status of all procurement-related activities. From that a quick review would indicate the general health of such efforts. Given that the owner, contractor, and subcontractors will all be procuring critical materials and equipment, including the Responsibility Code, further clarifies who is responsible to speed things up, in the case of a delay.

Evaporative Activities

An unusual type of activity that occurs on virtually every construction site is one where no crews are present, and no work needs to be done. Consider the time required for concrete to cure or paint to dry. The general rules-of-thumb for concrete to cure to the point where it can take construction loadings is 2 weeks. Paint will typically require 1 day drying before additional work can be done on that surface. This type of activity can be referred to as an "Evaporative Activity."

Depending on the nature of the project, these evaporative activities may have a larger or smaller impact. For a cast-in-place concrete, mid-rise building, the "Cure Slab" activity may be directly on the Critical Path. Anyone who has ever painted knows that high humidity may delay paint drying several days. The judgment of the scheduler should be used to determine if such activities are included in the schedule. If they are, then the Category of Work code could be used to designate this type of nonwork activity.

Weather

One of the greatest known unknowns in a project is the potential impact of weather on the construction. A complete review of United States federal government guide specifications conducted by Steen et al. in "Severe Weather Impact Analysis for Military Construction Projects," documented the extent to which atmospheric conditions are explicitly covered in the specifications. In many cases, the requirement is based on temperature. This is because the evaporation required to cure certain activities, such as concrete, will not take place unless the temperature is within a specific range. Often this temperature requirement extends to the ground beneath the work face. Such requirements ensure that proper bonding between new materials and existing ones can be developed. In other cases, the presence of precipitation will damage materials.

Even if contracts don't identify specific weather constraints, a good scheduler would not violate these constraints. For example, it would not be reasonable for a schedule to show pavement activities during the month of January in Chicago, Illinois. Those who live in regions with cold winters know that even asphalt patching material might not last the winter; the contractor will have to wait for spring. Contractors must base their construction plans on the typical weather for that season and consider the impacts of that weather on the planned construction.

For building projects, contractors often make their own weather inside a building. Readers of the CPM Tutor will be familiar with the sight of buildings with plastic in the place of windows. This keeps the precipitation out and allows the contractor to provide some heat within the building. To reflect the importance of weather on a project, many projects will include a planning milestone called "Exterior Closure" in their schedules. Knowing that the closure of the building will be delayed from August to October may not be a problem for a project in Saint Louis, Missouri, but it will certainly be for a project in Fairbanks, Alaska.

Conclusion

Creating an effective plan requires knowledge of both the mechanics of CPM and an understanding of how the schedule can be effectively used during the project. This chapter provided some important rules-of-thumb that many schedulers have learned the hard way. To begin, all major contractually required milestones, such as the project start and end dates, should be explicitly included in the schedule. The schedule should also include milestones and/or hammocks that illustrate contractual and planned project phasing. Including milestones and phasing will greatly enhance the value of the schedule to communicate overall project status.

Including appropriate calendars and holidays is needed so that planned dates make sense to those using schedule reports. Typically, one 5-day workweek and one 7-day calendar week should be used. Given the increased complexity of schedules with multiple calendars, using more than these two calendars is not always helpful.

The application of templates to organize patterns of repetitive activities is also a valuable technique. Consistent activity numbering and coding are helpful since they help team members more quickly understand and filter schedules with a set of repeated activities. To help illustrate the movement of different resources through the facility, the "crew chase" should be modeled using templates organized around type of work, crew responsibility, and work areas.

The allocation of expected earned value, or "Cost-Loading" schedules, is often required by contracts. Splitting the delivery and installation of capital equipment can facilitate prompt payment for equipment on site.

Finally, the scheduler must be aware of potential delays to the project caused by outside sources. If such delays can be anticipated, the scheduler should explicitly address these delays by adding new activities or milestones to the schedule. If delays actually occur, then external factors affecting changes to durations to these new activities, or changes to associated milestone dates, can be quickly understood to be the cause of a delay.

CHAPTER 6

Using the Plan

Planning is a prerequisite skill for most complex human endeavors. Leonard Bernstein, the famous composer and conductor, is quoted as saying "To achieve great things, two things are needed—a plan and not quite enough time." In the previous chapter, techniques to create effective construction plans were introduced. While the project plan may begin well enough, delays and impacts can reduce the time available to complete the project. The most common of these delays and impacts are described in this chapter. While construction managers often report that they are rarely given enough time to complete their projects, with the techniques described in this chapter, effective construction managers often achieve "great things!"

Progress Updates

Most contracts require that the initial schedule be submitted to the owner for review and approval. Following approval, the team uses the schedule to evaluate progress and anticipate potential delays. The schedule may also be used to predict the impact of scope changes requested by the owner. Once the project is complete, the schedule may be used as a record of progress in case there are any remaining issues to resolve.

There is one critical assumption in the use of a schedule for anything other than that first owner deliverable. That assumption is that the project never changes. If the project never changes, then that schedule can be used for all kinds of important tasks. The reader will, of course, recognize that this assumption is incorrect. All projects change over time and the longer since

Notice to Proceed, the more inaccurate that initial schedule will become.

Finding the meaning of Dwight Eisenhower's quote, "Plans are useless, but planning is indispensable," helps to put the role of that initial schedule in context. The initial schedule is just a plan. Over time that plan will no longer be a good model of the project. It is the process of planning schedule updates to make sure the schedule reflects the reality of the actual project that is "indespensible."

There are only three activities required for the planning process. First, update the current schedule to reflect anything that has changed since the last update. Second, compare the new current schedule with the previous schedule and adjust as needed. Third, repeat until complete.

The planning process occurs, typically, on a monthly basis. The scheduler will begin to update the schedule by talking with crew foreman working on the job during the previous month. The purpose of these conversations is to ask the people directly involved about the status of the work that was started, worked-on, and completed during the month. Armed with this information, the monthly progress meeting simply covers the topics identified in the following paragraphs.

Actual Start and Finish Date

The scheduler's most important report for evaluating progress is the list of activities that could have started or ended during the month. The report should be sorted to ensure that in-progress activities are on the top of the list. In-progress activities are those that already have an Actual Start Date from the previous month. Below in-progress activities are the list of those that could have begun. This will be the list of any activity whose Early Start Date fell within the reporting period. An additional padding of 10 to 20 workdays to the Early Start Date may also need to be added to catch any activities that may have started early. With this report the list of activities most likely to finish during the month will be followed by those most likely to start provide a common basis for progress discussions.

In most cases, CPM software can produce such a report directly; however, the scheduler should take care to understand the criteria used by the software to produce that report. The first time running any such report on a new version of CPM software, the scheduler should double-check the contents of the report with the information obtained by creating the list manually.

Each activity on the list is evaluated in order and the first decision to make is, "Did the activity start?" or, "Did the activity finish?" In either case, the date on which that start or finish took place can be written directly on the schedule report. If at all possible, the agreement on Actual Start and Actual Finish dates should be mutual between the scheduler and the owner's representative. If the date is not mutually agreed upon by both the owner and the contractor, then the daily reports, labor time cards, or jobsite photos for that part of the work should be reviewed by all parties and a more accurate date can be determined.

Updating Remaining Duration

The clearest expression of progress on an activity is one that begins and completes during a given progress period. Often, however, the scheduler does not have this luxury. For activities with an Actual Start Date but without an Actual Finish Date, a conversation needs to be had about activity progress. The potential controversy and difficulty of estimating an activity's progress can be simplified with an understanding of what is important, and not important to consider when evaluating progress.

There are two options for evaluating time-based progress of an activity. The first possibility is to determine the duration that has been expended. The second possibility is to determine the duration that remains. To determine which alternative is most helpful to understanding the progress on the project, three cases will be considered. In the first case, the crew is performing the work consistently with its estimated rate of production. In this case, there will be no difference in schedule calculation between planned and actual. As a result, either the expended or Remaining Duration will be relevant.

In the second case, the crew is performing the work at a lower productivity rate than expected. In this case, there will be a difference in schedule calculation between the expended and Remaining Duration. If the Expended Duration is used for the calculation (at the planned productivity rate), then the Remaining Duration (at the planned productivity rate) will underestimate the resulting actual duration. This is because there is more work left to be done than would have been expected at the planned productivity rate. As a result, when productivity drops, the measure of time-based progress that provides the most accurate result is an updated estimated of the Remaining Duration recalculated based on the actual productivity rate.

In the third case, the crew is performing the work at a greater productivity rate than expected. In this case, there will be a difference in schedule calculation between the expended or Remaining Duration. If the Expended Duration is used for the calculation (at the planned productivity rate), then the Remaining Duration (at the planned productivity rate), will overestimate the actual duration. This is because there is less work left to be done than would have been expected, at the planned productivity rate. As a result, when productivity is greater than expected, the measure of time-based progress that provides the most accurate result is an updated estimate of the Remaining Duration recalculated based on the actual productivity rate.

While the backward-looking Expended Duration may seem like a relevant measure, it is similar to the evaluation of "sunk cost" in evaluating capital expenditures. What has been spent, or sunk, in the past is not really relevant to determining the plan for the future. What is needed to best predict the future is an accurate estimate of the time remaining. Identifying the Remaining Duration of each in-progress activity is, therefore, the preferred method of measuring time-based progress.

Updating Earnings

On Cost-Loaded schedules, the total earnings on the project are also updated each progress period. This progress period is, typically every month. The monthly update process begins with the project team discussing the earnings for each activity. The sum of all earnings to date provides the total current earnings. The previous total current earnings are subtracted from this months total current earnings to determine the contractors' monthly progress payment.

With activities that begin and end during the same reporting period, the decision about the earnings achieved on that activity is clear. One-hundred percent of the value of activity could be earned. The same, however, cannot be said about activities that are in-progress. A monthly argument between the scheduler, contractor, and the owner's representative is not an effective project management technique to resolve differences in opinion about earnings, therefore the scheduler and contractor need to have a more effective way to justify monthly earnings.

The first way to smooth the discussion about earnings is to ensure that activities with a large equipment component be separated into two activities: one for the delivery of the equipment, and one for the installation of that equipment. The discussion is simplified because the contractor can be paid for on-site equipment once a paid invoice is provided. The contractor does not need to discuss the earnings of the equipment delivery versus the earnings associated with equipment installation.

The second way to smooth the discussion about earnings is to ensure that activities do not span multiple payment periods. If an activity begins in a previous period, it should be constructed so that it will complete during the later period. This is accomplished by setting activity durations to less than the monthly progress period. Even if there is difference of opinion about the expected earnings during the starting period, these differences will be resolved during the later period.

Separating time-based progress (in Remaining Duration) from earnings-based progress (in Percent Complete) has one more important implication. In many activities, it is possible that while the work of the activity is essentially complete, there are a few additional items to finish. In the United States, the list of remaining tasks is called a Punch List. To reflect this Punch List on an otherwise completed activity is straightforward. The activity will have an Actual Start Date and an Actual Finish Date to allow following work to proceed. The activity will also have a Percent Complete equal to the planned earnings less the value to repair the Punch List. When the Punch List items are all completed, full earnings can be achieved.

Updating Sequence

The need to update plans is, of course, well known. A Roman slave in the fifth century BCE, Publis Syrus, quipped that, "It is a bad plan that admits of no modification." Depending on the way the project proceeds, the sequence of the schedule may also need to be changed to reflect the conditions on-the-ground.

Since any change in sequence could have significant impacts upon the overall project plan, contractor-driven changes in sequence are best described with a one time schedule change submittal that includes: reason for the change, list of activities affected, before- and after-change schedule calculation, and a statement that the schedule is not adversely affected by the needed change. Taking care of sequence changes as a submittal ensures that a clear discussion of the reason, and the marginal impact of change can be identified, discussed, and acknowledged by the owner or owner's representative.

Out-of-Sequence Progress

An aspect of updating CPM schedules that often occurs is called Out-of-Sequence Progress. To explain the importance and resolution of Out-of-Sequence Progress, consider the schedule shown in Table 6.1. This schedule contains three sequential activities, "Dig Trench," "Place Culvert," and "Backfill." Three separate crews are expected to perform this project. The first has excavating and surveying equipment to dig the trench at the correct location

Activity	Duration (Workdays)	Preceding Activity
Dig Trench	20	None
Place Culvert	10	Dig Trench
Backfill	15	Place Culvert

TABLE 6.1 Culvert Project Schedule

and depth. The second crew has the needed trucks, cranes, and equipment to piece together the culvert as it arrives on-site. The third crew comes behind the culvert crew and backfills and compacts the soil around the culvert. Figure 6.1 shows the Fenced-Bar Chart for the example project. Figure 6.2 shows the project in an Arrow Diagram with completed CPM calculations.

On day 15 of the project, the scheduler collects progress information. For each activity that is in-progress, the Actual Start Date of the activity and the Remaining Duration are determined. The right three columns of Table 6.2 provide this information. Notice that Out-of-Sequence progress

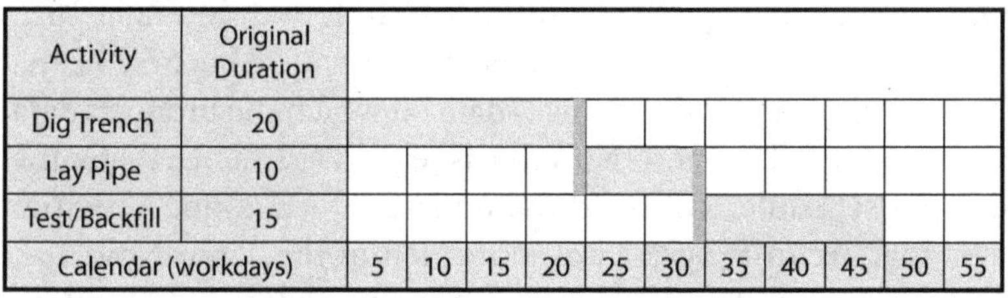

Figure 6.1 Culvert Fenced-Bar Chart.

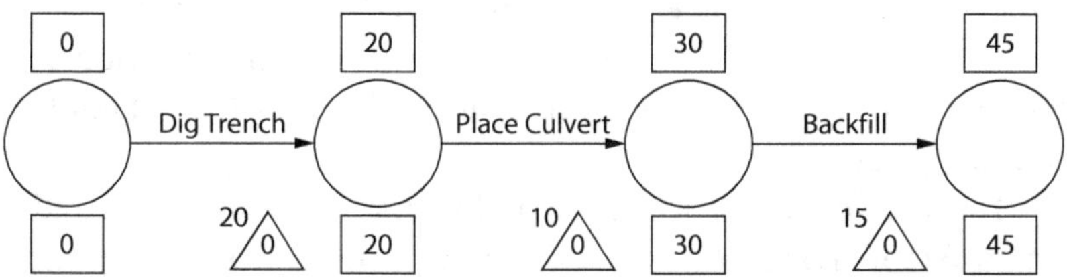

Figure 6.2 Culvert Arrow Diagram.

Activity	**Duration (Workdays)**	**Preceding Activity**	**Actual Start Date**	**Expended Duration (Workdays)**	**Remaining Duration (Workdays)**
Dig Trench	20	None	1	15	5
Place Culvert	10	Dig Trench	12	3	10
Backfill	15	Place Culvert	14	1	15

Table 6.2 Out-of-Sequence Schedule

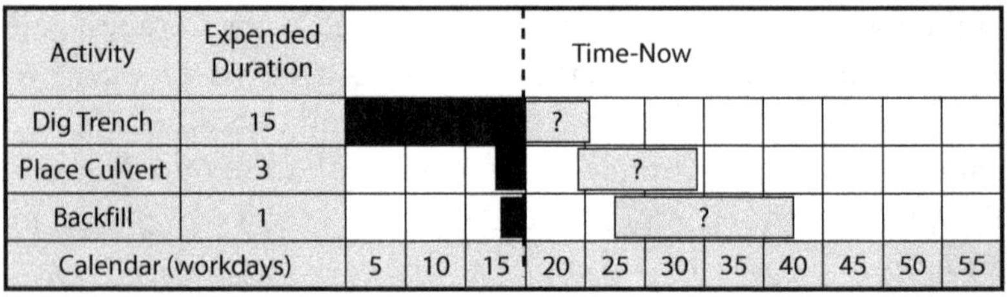

Figure 6.3 Out-of-Sequence Fenced-Bar Chart.

has taken place. All three activities have started, but the preceding activities are not yet completed.

Figure 6.3 updates the Culvert project bar chart with the Out-of-Sequence progress information found in Table 6.2. Figure 6.3 introduces several new elements. The first is the "Time-Now" line. Time-Now is date on which the progress data was collected. The Time-Now line is set to day 15 since that is the day the scheduler collected the update information. Activities bars that are in-progress are shown left of the Time-Now line. The second additional element is that the Expended Duration of each started activity is shown shaded, left of the Time-Now line. The third new element, the Remaining Duration is also added to the chart. The Remaining Duration of each activity is shown to the right of the Time-Now line—the question is where to show those bars given that the logic of the schedule has now been broken?

The Out-of-Sequence information has been also updated on an Arrow Diagram in Fig. 6.4. Each activity arrow that has started has been time-scaled to show its Remaining Duration. The difficulty of providing a calculation, at this point, is to know

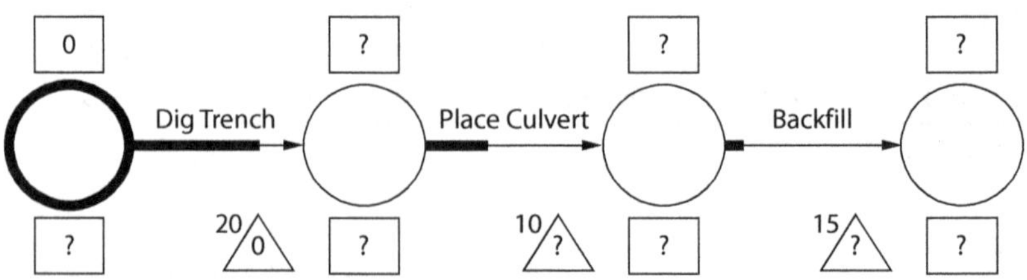

Figure 6.4 Out-of-Sequence Arrow Diagram.

how to proceed, given information that directly contradicts the CPM rules. In CPM, an activity may only start when all prior activities have been completed.

Out-of-Sequence progress is a demonstration of one way that CPM, or any discrete model of a semi-fluid construction process, cannot take into account all aspects of a construction project. No matter how hard the scheduler tries, a perfect schedule cannot be created. This limitation should not, however, be taken as a recommendation to begin to experiment with other types of construction models for two important reasons. First, CPM is required by contract so it is in the interest of all parties to make it work as well as possible. Second, no matter what additional type of modeling is done, none will ever provide a 100% match between the schedule and the job. Regardless of the latest "great new" idea in project planning promoted by an organization that wants a donation to find out, or whatever is required by one owner or another, there is simply no project planning method that can predict the uncertain construction process. All project plans are simply models of a messy construction process. As Dr. Box has so clearly stated, "All models are wrong. Some models are useful." The goal of the CPM tutor is to highlight problems associated with Out-of-Sequence progress so that rather than dismiss CPM, the scheduler will know how to make it "useful" even when it is "wrong."

CPM Progress Calculations

The first consideration when calculating a schedule with progress is to identify the date from which schedule calculation is to be run. The name of this schedule calculation date differs among scheduling software. The CPM Tutor refers to this date as the "Time-Now." Typically, Time-Now is not the same day as that on which the schedule is actually calculated. This is because the scheduler will collect the in-progress updates for each activity and collate that entire set of information several days before running the schedule.

The next consideration for evaluating time-based progress is that completed activities are no longer considered in schedule calculations. Scheduling software should mandate that if an

Actual Finish Date is posted, then the Remaining Duration of that activity is automatically set to zero.

Processing in-progress activities is the next step. Once an Actual Start Date is provided, the Early Finish Date of the activity is derived from the Time-Now plus the Remaining Duration of the activity. If no value for Remaining Duration has been entered by the scheduler, then the Original Duration should be automatically set to the Remaining Duration.

As introduced in the previous section, only real concern when calculating in-progress schedules is in the case of an activity that has begun without having the prior activity being finished. This Out-of-Sequence progress cannot simply be handled as other in-progress activities. Fig. 6.4 illustrated a three-activity schedule with Out-of-Sequence progress. If the standard in-progress calculation method was used to calculate the schedule, the result could be as shown in Fig. 6.5. Each activity's Early Finish Date would simply be calculated by adding the Remaining Duration to the Time-Now date. Notice that this approach completely disregards the schedules original logic.

An alternative to disregarding the schedule's logic is to allow Out-of-Sequence activities to be calculated based on the Remaining Duration of in-progress activities and retain the logical sequence provided by CPM. Figure 6.6 illustrates how the two following activities in Fig. 6.5 are properly scheduled according to their prior activities scheduled completion. The assumption that the logic still applies to the Remaining Duration of the activities is the more conservative, and hence the preferred method of scheduling Out-of-Sequence activities. The scheduler should never disregard the logic of the approved schedule (even

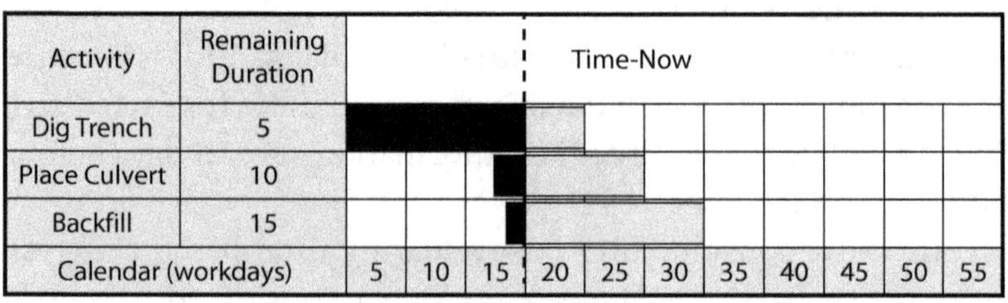

FIGURE 6.5 Out-of-Sequence—disregarding logic.

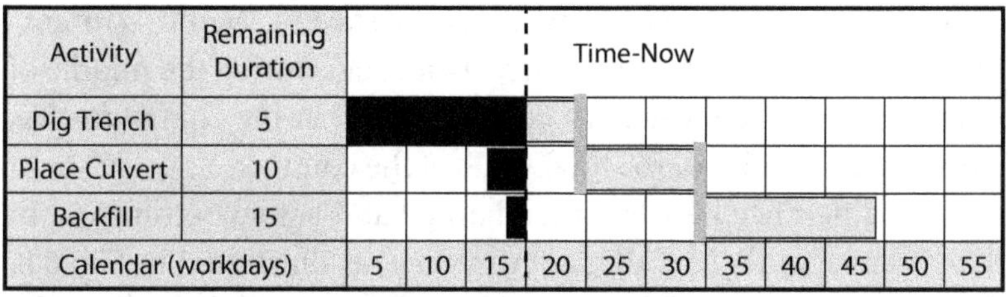

FIGURE 6.6 Out-of-Sequence—maintaining logic.

if that option is provided in scheduling software). If the logic is incorrect, it can be changed, but the change must be explicit and agreed upon, not hidden in a scheduling software setting.

Earned Value Management

Earned Value Management (EVM) is an involved technique required on very large, multi-year, cost-reimbursable projects. The use of EVM on fixed-priced construction projects focuses on only one type of analysis that can be simply summarized as "compare planned to actual." Figure 6.7 provides a motivating example. In this example, both planned and actual earnings are shown by their characteristic S-curve. The solid line is the planned earnings, derived from the originally approved, or updated schedule. Such a schedule is often called a "baseline schedule." At each reporting period, the cumulative earnings have been posted on the dotted line. To get ready for

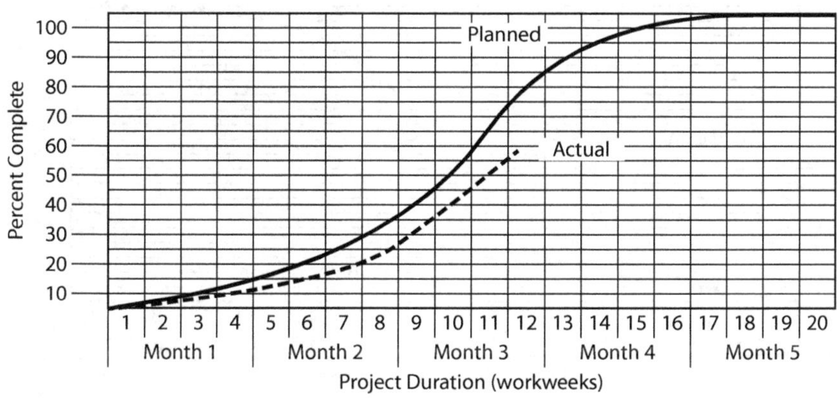

FIGURE 6.7 Planned versus actual earnings.

the contractor to submit an invoice at the start of "Month 4," information is collected about the current earnings during the middle of week 12. The actual earnings curve shows that the contractor has placed at about 65% of the total value of the contract.

The difference between the planned and actual earnings is, in the parlance of EVM, called a "Variance" as illustrated in Fig. 6.8. There are two types of Variance for any given schedule update. The first is the "Cost Variance." As noted previously, "cost" in EVM (in the context being discussed here) is actually the cost to the owner, which is, in the context of the CPM Tutor the contractor's earnings.

Cost Variance is the difference between the Budget Cost of Work Scheduled (BCWS) and the Actual Cost of Work Performed (ACWP). The BCWS is 85%. The ACWP is 65%. Subtracting the planed progress, or BCWS, from the actual progress, or ACWP, is −20%. A negative Cost Variance indicates that the work is 20% behind the expected earnings at this reporting period.

The next type of Variance of interest to EVM is the "Time Variance." Time Variance is the difference between the planned duration, at the current level of earnings, versus the actual duration. In the example from Fig. 6.8, the BCWS was to achieve the current 65% progress at the end of week 10. The ACWS is actually calculated based on the information recorded during the middle of week 12 (estimated to be week 12.5). Subtracting the durations shows that the project is 2.5 weeks behind schedule.

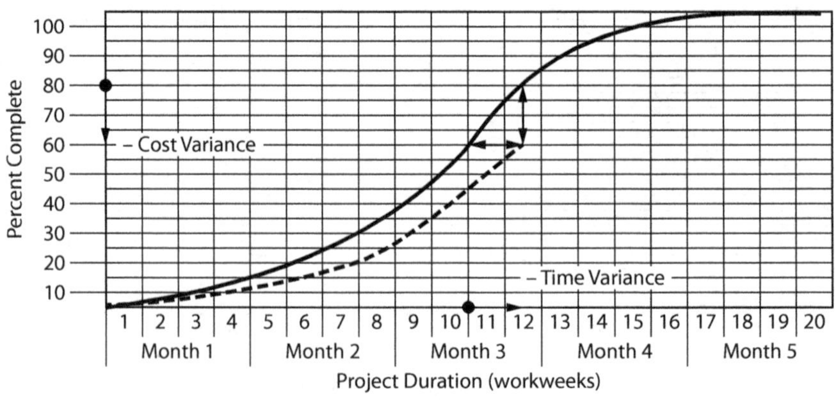

FIGURE 6.8 Planned versus actual—negative variance.

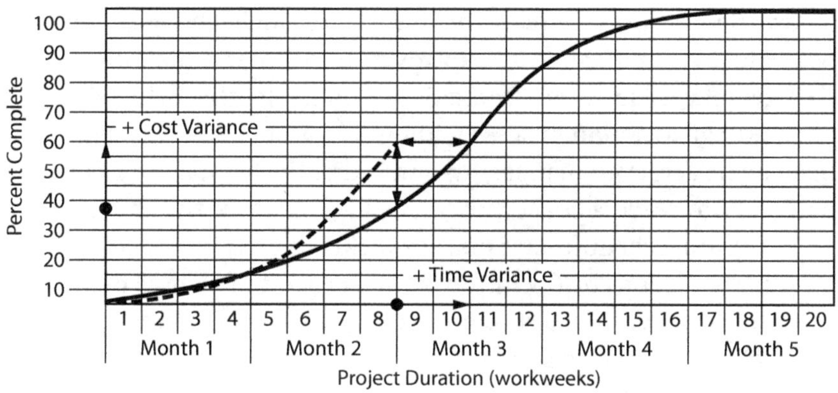

FIGURE 6.9 Planned versus actual—positive variance.

The Cost Variance and Time Variance values are positive for schedules that are ahead of schedule. Figure 6.9 shows the planned versus actual earnings curves for a different project. The planned earnings curve is shown as a solid line. The actual earnings are shown as a dotted line. At the end of week 8 when the progress information was collected and posted to the schedule, the project had an ACWP of 55%. If the contractor had been performing work per the schedule, the BCWS would have been about 32.5%. Since the ACWP is greater than the BCWS, the project is ahead of schedule with a Cost Variance of positive 22.5%. Similarly, the Time Variance is a positive 25% meaning that after 8 weeks the project is 2 weeks ahead of schedule.

Change Analysis

As was noted earlier in this chapter, the initial plan for a project will always change over time. For some projects, the plan will change only slightly. For other projects, the original plan may substantially change by the time the project concludes. As noted at the start of this chapter, the process of planning begins with agreement on an initial schedule.

The mechanics needed to update the schedule are always the same, regardless of the nature of the change. Activities are added, deleted, or changed as is the sequence between them. In the best situation, these changes are evaluated and submitted before being encountered. The impact of such changes can be evaluated with the least controversy. In other situations, a change may require an immediate update to the originally planned construction process without prior approval. The best that can be achieved in that situation is that the team takes steps necessary to reach a consensus about the impact of the change, and resulting schedule changes, as quickly as possible.

The paragraphs below introduce the process of change management using CPM. At its core, the objective of this process is not really about the schedule. The schedule is just the tool. The objective of the process is to assess the cause or effect of a change and to predict the resultant impact on the total cost and duration of a project. As has been noted, such a process is "indispensable."

Maintain a Baseline

The analysis of project changes is contingent upon having a baseline schedule that includes all significant features of work, identifies the subcontractor or partner responsible for the completion of the tasks, identifies nonwork activities, and shows work areas through which the crews must complete their efforts. The schedule will have had to be updated during the project to accurately reflect progress to date. Over time, there may be several baseline schedules. Each later baseline reflects previously agreed upon changes to that point in the project. If the schedule does not have an accurate baseline, the results of the analysis may not be agreed upon.

Identify the Critical Path

The Critical Path Method, by definition, identifies the sets of activities that govern the overall duration of the project. As a result, the first step in the evaluation of any delay is to identify the network paths that lead to and follow from the current schedule snapshot. This is best accomplished using two specific reports developed from the CPM scheduling software. The specific names of these reports may differ with each scheduling software product; however, generically these reports can be called the "Total Float Report" and the "Activity Logic Report."

The Total Float Report lists activities in ascending order of their Total Float. The subsort for this report should an ascending Early Start Date. Those activities on the Critical Path will all be sorted to the top of the report and then listed according to when they could begin. Remember that if there are intermediate millstones in a schedule, then a set of activities on a given path need not have the same Total Float value. Activities closer to the Time-Now date will appear at the top of the list and activities that complete the schedule will follow.

Large schedules often have multiple Critical Paths. Thus, the Total Float Report alone is insufficient to evaluate the change impacts. Another report, the Activity Logic Report, is also required. The Activity Logic Report will show each activity and its preceding and/or following activities. This report will list all activities, including completed activities, typically sorted by Activity ID. With this report, the scheduler can determine the set of activities on each Critical Path. Completed activities that also have contributed to any current delays will also be shown on the Activity Logic Report.

List Delayed Activities

There are three situations in which a CPM schedule is delayed. The first is that an activity on the Critical Path started later than its Late Start Date and completed as scheduled. The late start of this activity will push back all subsequent activities along the Critical Path. The second situation is where an activity starts on time and takes longer than expected. In this case, the delayed Actual Finish Date of the activity will push back all subsequent activities long the Critical Path. The third situation is where there is a late start and extended duration.

To identify late and delayed activities from the past, the scheduler will have to return to previous schedules and review the Total Float and Activity Logic Reports. At this point in the process, the "Forensic Scheduler" is interested in cataloging every activity affected without regard to the cause. The tasks to be accomplished include, but are not limited to, the following:

1. List all activities for which the Actual Start Date occurred later than the previously planned Late Start Date.

2. List all additional activities for which the Actual Finish Date occurred later than the previously planned Late Finish Date.
3. List all additional activities for which the Actual Duration exceeded their Original Duration.

Once the list of delayed activities has been identified, the following analysis is to be conducted:

1. Was the impact on a single Critical Path, multiple Critical Paths, or a non-Critical Path that became critical as a result of the delay?
2. Was the delay the result of a single point that pushed back other activities, or were there multiple delays?

Determine the Cause

With the list of delayed activities and an understanding of the nature of the delay, the scheduler can begin to search for causes. Ideally, the scheduler will have access to on-site personnel, daily reports, jobsite photographs, requests for changes, etc. These sources should be reviewed once the delayed activities and the date ranges for the delay(s) have been determined. When reviewing these sources, the scheduler may consider any of the following types of questions:

- Weather delays may be identified by delays of similar activities along several paths through the schedule during shared, discrete periods of time.
- Material procurement delays may be identified by a one-time occurrence, several delays with the same type of material, or a delay with all procurement leading up to a specific activity.
- Productivity delays of prime contractor activities may be identified by determining the projected man-hours per day to man-hours spent each day, provided that such information is available from the contractor's cost estimate.
- Productivity delays may be identified by a common delay of activities that share the same Responsibility Activity Code.

- Site access delays may be identified by a delay that occurred to a set of activities delayed for a similar time frame.
- Delays associated with specific changes to scopes of work or site conditions must be verified with the Contractor Quality Control Officer and/or Superintendent.
- Delays associated with rework should be evaluated, in consultation with the Contractor Quality Control Officer and/or Superintendent, to determine whether that rework was directed by the owner's representative.

Determine the Pattern

The scheduler should provide a succinct description of the cause of all delays found. The scheduler should be careful to individually identify concurrent delays. A very important aspect of this description is the determination of the delay as being a delay that occurred in the past, or an ongoing delay. To resolve an on-going delay, some changes must be made to future activities. Therefore, the scheduler must identify the activities likely to be affected if the cause of the delay is not resolved. The patterns of delay found may, for example, be productivity delays that pertain to activities in a given work area or of a given crew type.

Predict the Full Impact

Provided that the contract completion date was properly assigned a Finish-No-Later-Than milestone date, the delayed schedule's Critical Path will have a negative Total Float. Only the lawyers can decide who is at fault and who is not, as well as who gets compensated and who does not get compensated. The job of the scheduler is not to fight that legal battle, but to "press on" and get the schedule back on track. The scheduler must do something to try to mitigate the delay and make up that negative Total Float.

Often the answer is to speed up activities on the Critical Path. If a schedule is 7 working days behind, then the scheduler may try to make up that time on the activities that are in-progress or have not started yet. Unfortunately, once a project is behind schedule, it may not be enough simply to make up for lost time. That is because the impact of the delay, once the cause is known, may affect future activities. Before deciding on a course of action, the scheduler must predict the full impact of all delays.

A phenomenon widely known by all who have spent any time on a construction project is that once a project gets behind schedule, it tends to become further and further behind schedule. When a delay is encountered, the project team may try to speed up activities on the Critical Path to make up that delay. For example, Fig. 6.8 shows an actual schedule that has a Negative Time Variance. The intuition of the Earned Value approach would lead a novice scheduler to believe that if the earnings curve can be left-shifted, then the schedule can be put back on track. The possibly incorrect assessment of this delay is shown in Fig. 6.10.

It is often assumed that resolving a past delay will give a good estimate of future progress. This is not true if the delay impact is ongoing. If the delay is ongoing, then the Actual Durations of future activities will be greater than their Original Durations. If these delayed activities are on the Critical Path, then the schedule will get pushed even further back. This delay will occur even if a few activities are accelerated to make up for the original negative float. Figure 6.11 provides the one possible illustration that the shape of the projected ACWP curve actually changes.

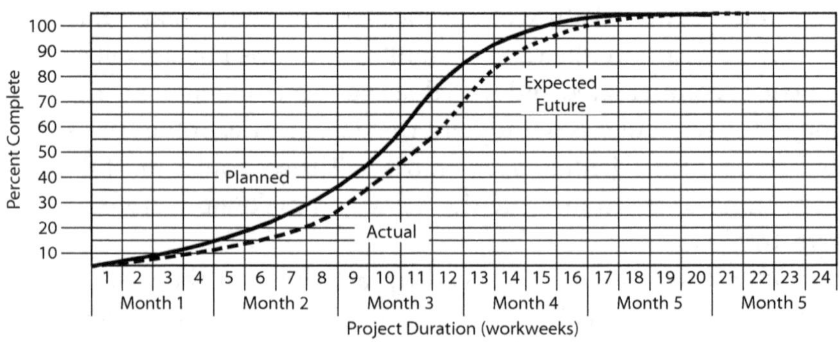

FIGURE 6.10 Expected future delay.

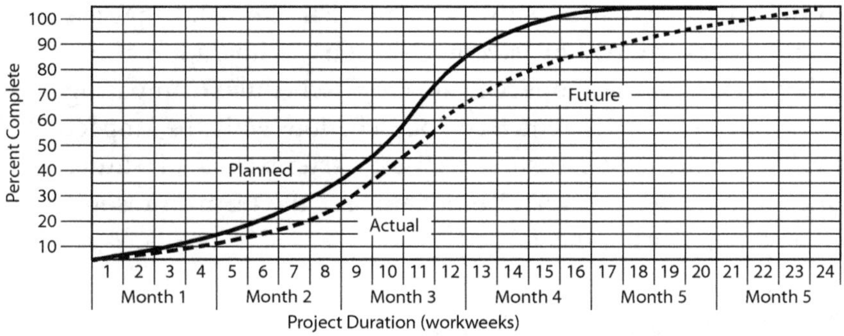

Figure 6.11 Real impact of compounding delays.

To turn CPM into an effective crystal ball, the patterns of delay that caused problems in the past must be applied to future activities. These activities will have their Original Durations increased to reflect the "real future" scenario. To evaluate the possible future patterns of delay, consider finding answers to questions such as those in the list below. The scheduler should ask the question for each group of similarly delayed activities. As there may be several concurrent delays, the scheduler will have to use best judgment about which delays to consider for each category of delayed activities.

- What is the rate of Total Float loss?
- What is the difference between Actual Duration and Original Duration?
- Do the rates of delay apply to the contractor's crew or subcontractor activities?
- Do the rates of delay apply to all work in a specific work area?
- Do the rates of delay apply to all work of a given type?

For large schedules, the answers to these questions may not be obvious. If there is information available on the progress of several hundred activities, then the answers may be best found by transferring the schedule data to spreadsheets or databases for statistical analysis. If the answers to these questions can be found and supported with data, then the scheduler has to find those activities with the same characteristics that are currently in-progress or have not yet started, and apply those delay factors in an "Impact Evaluation" schedule. Comparing the results of the current Baseline Schedule with the Impact Evaluation Schedule, using realistic assumptions of future impacts, can be a powerful motivator for the team to resolve whatever is causing the delay.

Develop a Recovery or Mitigation Plan

No manager ever wants to be presented with a problem without also being given a recommendation to resolve or mitigate the problem. In this stage of planning, the scheduler considers alternatives to the problems identified, and with the time remaining in the project, develops one of two types of plans. The first type of plan, a "Recovery Plan," is only possible if the delay is caught early enough to resolve the underlying cause of the problem on all future activities. The second type of plan, a "Mitigation Plan," is one in which the project team knows there will be an unavoidable impact on the project, but together seeks to minimize the impact of that plan.

Knowing the cause of the problem and its expected impact is essential to developing either a Recovery Plan or a Mitigation Plan. Schedulers will ask one or more of the following types of questions to discover opportunities to develop their plans:

- Could more resources be added to activities with durations over 2 weeks (or other long duration activities)?
- Could sequential Critical Path activities be performed concurrently?
- Could some work on the Critical Path be shortened by adding more resources?
- Do activities of similar type show consistent productivity rates?
- Do we accelerate prime contractor work before evaluating subcontract work?

Construction Claims

Projects proceed through a series of major phases: planning, design, build, and operations/maintenance. A phase that is often overlooked is pejoratively called the "claims phase." This phase typically begins just before the end of construction and may extend a decade passed the completion of construction. In most cases, the claims phase occurs because the project team is unable agree on the impacts of changes encountered during the project.

Like any other project phase, effective management of this phase of the project demands that the scheduler understand the scope of the work to be accomplished and the means and methods needed to complete that work. The management of the claims begins with a working knowledge of what constitutes a claim. The objective of this section is not to serve as a legal text or guide. The objective is to provide general knowledge of construction claims from the point of view of how schedules may be used as part of the claims process. Readers with specific questions should consult with an attorney.

What Is a Claim?

While the succinct answer is, "A claim is a request for compensation resulting from differences between what was in the contract and what was accomplished on the project," this statement needs some explanation to be of practical use. The basis of any claim is the contract. Without a contract, there is no basis for a claim. Thus, subcontractors may not submit a claim to construction owners. Subcontractors' contracts are with the prime contractor, not the owner. However, claims submitted by subcontractors may be submitted by the prime contractor.

The content of the claim is "compensation." In the context of construction contracts, there are only two types of compensation. Adjusting the total contract amount (either higher or lower) is the first type of compensation. The other type of compensation provided to construction contractors is that of changes to the duration of the construction contract (either shorter or longer).

Notice that the content of the claim is not about "fairness" or "justice." The short tempers and high blood pressure often associated with construction claim "discussions" cannot soothe the conscience or teach a lesson. The only compensation offered in the construction claim is time and/or money.

The definition of a claim makes no statement about the chances that the owner will provide the requested compensation. Several hurdles have to be jumped during the claims analysis process to determine whether compensation will be provided. The first of these hurdles is to determine whether the owner is responsible, under the terms of the contract, to provide any compensation at all. There are many cases where the owner is not required to provide compensation. In other cases, the contract limits the types of compensation that can be provided. There are three general categories of delays into which claims fall. These are described in the following sections.

Compensable Delays

Compensable delays are those caused by the owner alone. These are deviations between the original contract and the work performed for which the contractor was not compensated. It is ususally when these changes affect activities on the Critical Path that a compensable delay occurs. Consider the example of an owner who wants additional work done in a specific part of the project. Further consider that the owner pays the contractor for this additional work but does not extend the duration of the project in that contract modification. The contractor's subsequent claim for additional time resulting from that change will hinge upon the placement of the changed activities in the schedule. If the work was on the Critical Path, then the owner is, most likely, responsible for a schedule delay and the time is "compensable." If the work is not on the Critical Path, then the contractor will be paid for the change to project scope, but be less likely to receive additional time. The typical categories of compensable delays are described in the following paragraphs.

"Differing Site Condition" claims are based on differences between the contract documents and the actual conditions on this site. Such claims commonly occur when subsurface conditions are different than appear on the construction contracts. Other claims could occur when working within existing buildings and hazardous materials, not identified in the original contract, are found. Another type of Differing Site Condition claim occurs when a contractor bases a bid on some specific means and methods, only to find that site conditions require a less efficient method of construction.

"Acceleration" claims occur when additional work is added to the contract without providing additional time. Thus, the original contract work must be "accelerated" to ensure that the contract completion will not be pushed back. The contractor will have to produce very clear records showing that the owner directed work to be added or accomplished in a way that directly added or changed specific construction activities and their sequence. Acceleration claims may also include reduced productivities caused by inefficiencies in how such work was directed.

"Inefficiency" claims are a broad category of claims that include, but are not limited to, unexpected restrictions to job site access, limitations to material staging areas, and delays to shop drawing approvals. In many cases, owners will specifically discuss these topics in the bidding documents. Since such impacts are clearly stated in the contract, contractor bid documents should take these factors into account prior to bidding the job (and not during the claims phase). If such impacts are not in the contract, the contractor's strongest position will be to have records of both worktime and nonwork hours caused by such delays.

"Project Impact" claims are all-inclusive claims, often the result of an accountant's post-completion financial analysis rather than an analysis of specific instances of delays on the project. These are often called "get well" claims since the objective is to try to squeeze the owner for additional funds, allowing the project to a achieve specific fiscal objective. These claims are very difficult to evaluate and resolve since they lump multiple delays into a single claim. Such claims may or may not be compensable depending on the subanalysis of the individual delays.

Compensable claims may be the result of the owner's actions. In acceleration claims, for example, the owner or owner's representative directs the contractor to change the contract, but not delay the overall contract completion. Documentation on such claims requires the contractor to provide evidence of the direction to make these changes. The contractor should be very careful not to follow such direction without written documentation.

Compensable claims may also be the result of the owner's inaction. For example, if the owner provides late shop drawings review and that delays procurement and installation of a system, then the owner's inaction may be identified as the cause of the problem. Site access problems can occur because the owner's representative fails to properly coordinate with the facility security office to allow site access.

Noncompensable Delays

The categories of compensable delays and their resolution process are often explicitly defined by the contract. Including this information in a contract allows the parties to more clearly understand the type of risk they are assuming by signing that contract. Consider a generic contract without a changes clause. The risk of such a contract depends directly on the type of project. If the project is a utility replacement in the central business district of a major city there will be many undocumented utilities and conditions different from that shown on contract documents. Without a changes clause, construction company insurers on such a project would not provide a performance bond. The contract's changes clause reflects the fact that the full set of all problems that will arise on any one project cannot be known ahead of time.

The governing theory of such changes clauses is that risks are apportioned to the contracting party best able to manage that risk. Many types of problems could arise on a construction project that are the responsibility of the construction contractor. The contractor's home office staff, for example, may forget to submit purchase orders for key materials. In another example, the labor market could change between the time of the bid and time of the construction, driving up labor costs. A contractor, in another example, may be unable to rent a key piece of construction equipment when needed, delaying the start of a Critical Path activity. The risk of resolving delays caused by the contractor's or subcontractors' actions (or inaction) with regard to the execution of the work are assigned to the contractor. As a result, these types of delays are not reimbursed by the owner. Such delays are called "Noncompensable Delays."

Excusable Delays

Contracts allocate some risks to the owner and others to the contractor. There is another category of risk that cannot be apportioned to either party. Delays associated with this type of risk are called "Excusable Delays." An important factor in evaluating excusable delays is that both parties are required to consider "normally expected" delay as part of their plans. For example, a large building project that breaks ground in the upper Mid-Western United States in October can be expected to encounter major weather delays to completing foundation work due to normally expected cold temperatures and precipitation. Excusable delays for out-of-the-expected conditions are often resolved by providing contract time extensions without compensation for costs.

Claims Analysis

The following paragraphs describe the established process by which CPM schedules are used to support claims analysis. This process has been established by its repeated use in court cases. Failure to follow this procedure has resulted in the CPM schedule being thrown out as evidence in legal proceedings. The objective of this established process is to create an objective basis for the identification of time-based compensation, i.e., contract time extensions. Even if the analysis is conducted by one party or another in the dispute, having this objective allows a precise discussion of specific delays.

Recover the "As-Planned Schedule"

The "As-Planned Schedule" is the schedule that was submitted by the contractor and approved by the owner. Some contracts require the contractor to provide an "Initial Schedule." The Initial Schedule may not be the one most appropriate for use as the As-Planned Schedule. This is because the Initial Schedule often covers only the tasks needed for the contractor's mobilization phase. The As-Planned Schedule recovered from the project records should be the first schedule approved by the owner that covers the entire project. This schedule will reflect the contractor's plan for completing the work.

Recover the Monthly "As-Built Schedules"

The "As-Built Schedules" are those monthly schedules submitted by the contractor and approved by the owner. The As-Built Schedules show the time-based and earnings progress of the work over time. Daily reports and jobsite photographs should confirm that the progress on the schedule reflects the reality at the workface. The As-Built Schedules must also incorporate updates to reflect all mutually agreed upon changes. If As-Built Schedules were not used to document progress payments and updated to reflect mutually agreed upon changes, then the use of CPM to support the claim may be thrown out entirely.

The scheduler should recover the As-Built Schedule up to the month before the claimed situations occurred. It may also be necessary to update the schedule with interim progress information up to the day before the claimed situation occurred.

Develop the "Would-Have-Been Schedule"

In the initial definition of a claim, the compensation associated with a claim was based on the difference between what was in the contract and what was expected. In the next step of claims analysis, the As-Built Schedule at the time of the claim is projected into the future to predict the expected progress if the claim had not occurred. The Would-Have-Been Schedule will need to be clear and understandable to all parties to be accepted as part of the legal proceedings.

Reclaim the "As-Built Schedules"

The next step in the claims analysis process is to obtain all schedules that follow the claimed situation. These schedules, if used to assess time and earnings progress, should be able to demonstrate the actual impact of the claim on the project. The specific causes and impacts of each concurrent delay should be identified activity by activity. The process is the same as the previously described delay analysis process.

The contractor should be careful to ensure that compensable, noncompensable, and excusable delays are all included in these As-Built Schedules. Failure to have included all actual delays will cast the use of CPM as other than an objective tool. It is possible, but time consuming, to construct As-Built Schedules from project records; however, their validity may be seen as less than objective.

Compare and Analyze

The previous steps of the claims analysis process have been getting ready to compare the planned and actual project execution. Like many other types of construction activities, getting ready to do the work is often more time consuming than actually doing the work. At this stage, the Would-Have-Been Schedule is compared to the As-Built Schedules to list each and every delayed activity and document the cause for that delay. In the analysis, the basis for determining the delay cause as compensable, noncompensable, or excusable is included.

During the process of recovering and/or recreating the As-Planned, Would-Have-Been, and As-Built Schedules, the claims analyst will have developed much of the justification needed in this stage of the process. The job of the claims analyst is to compile that information into a truthful and compelling narrative.

The narrative prepared here is, in many ways, more important than the original claim. Without the analysis, there is no objective explanation for the requested compensation. The prudent contractor wishing to quickly resolve the claim should take the time to develop and submit this analysis with their claim. If that were accomplished, man-years of effort, and legal fees, could be eliminated from the claims phase of every large project.

Compensation

The cost-estimating techniques needed to determine the direct cost of additional work are the subject of cost engineering texts and outside the scope of a scheduling text. From the view point of the CPM schedule, the compensation of most interest to the scheduler is the additional time required to complete the project. The costs of interest to the scheduler are primarily the costs that follow from that additional time on the project; these are indirect costs.

The first indirect cost, that of field office and field supervision, is relatively easy to calculate when work is added to the project. This calculation should be made based on the actual cost to rent the field office trailer, provide utilities, and project-specific vehicles or equipment. Additional costs to maintain site personnel during the delay can also be calculated. Work stoppages are more difficult to calculate since in a complete job shut-down, all personnel may be removed from the site, but contracts trailer rental or other costs may still be in effect. In the case of partial shutdowns or limitations of access, more detailed analysis will need to be conducted. Regardless of the specifics of the change, the impact on the field office should be a matter of obtaining rental payments, utility bills, and other project-specific documentation that pertains to the claim.

The second type of indirect cost, that of home office overhead, is much more difficult to calculate. This is because most contractors simultaneously perform projects of different scopes and durations. Only part of the wages and expenses of the home office can be allocated to a given project. The justification for home office overheads is that if the project is delayed, the contractor has to allocate additional funds to keep home office personnel working on the subject contract. The assumption is that if they were not working on the delayed project, they would be working on new projects. Proportional allocation of home office costs is often allowed since home office personnel are not allocated to indivdiual projects but funded through pooled project overhead accounts.

Claims Analysis Conclusion

A CPM schedule is a very helpful tool for claims analysis. The use of CPM in claims analysis is to objectively identify the impact of the claimed work on the project's finish date. Just because CPM can be used for claims analysis does not mean that the scheduler should encourage construction claims. On the contrary- the scheduler's first choice should be to head off claims entirely by negotiating all change orders, to include full compensation for both time and cost, during the project. One of the most thankless jobs in all of construction management is to have to reconstruct the schedule from daily reports and progress photos after the fact.

CHAPTER 7

Advanced Topics

ADM and PDM begin to answer several important questions about construction projects such as when should work crews begin and end work and what will the expected earnings be at a given point in the project. This chapter presents techniques that can be used to develop alternative resourcing approaches that consider contractor profits versus simply overall project earnings.

To evaluate shortages in critical resources, the "Limited Resource Allocation" technique is introduced. If, on the other hand, there are sufficient resources, then the scheduler will want to ensure that crews can come on the site, work through the project, and move to the next project without interruption. To reduce worker turnover, this chapter presents the "Resource Leveling" technique.

At this point in the book, the scheduler can predict earnings based on assumed productivity rates, but cannot assess the impact on the contractor's actual costs. In this chapter, two cost-based techniques will be introduced. The first technique, the "Time-Cost Tradeoff," allows the scheduler to evaluate the impact of adding resources to schedule activities to try to reduce the duration of the project. The second technique, "Cash-Flow Analysis," allows the contractor to evaluate how project financing, payment terms, and subcontracting affects profitability.

The approach taken to these topics by the CPM Tutor is decidedly heuristic, or rule-based. While mathematical techniques can be applied to solve some of the problems described in this chapter, such methods are often impractical in practice. The scheduler should pay particular attention to the automated solutions provided by scheduling software related to the topics presented in this chapter. In many cases, the algorithms included in such software are not fully documented and may not be relied upon to provide an optimal answer. In such cases, the scheduler may spend more time trying to reverse-engineer the software algorithm than if they had conducted their own analysis.

The heuristic approaches described in this chapter are easily understood. This helps the scheduler communicate the results to the team. By combining alternative heuristics, the scheduler can obtain what is ultimately needed, which are options. Options are important since no computer model of a construction project will contain all the constraints found on the real jobsite. In the final analysis, the project team is responsible for the results on the project, not the scheduling software.

Resources Types

Before discussing methods for evaluating resource scarcity and efficiency, it is important to understand what is meant by the term "Resources." Resources are the means of production needed to complete a project. The three types of resources that typically require attention are labor, material, and equipment.

Labor

On large construction projects, labor is organized into crews that can perform specific types of work. It is typically assumed that all crews of a given type have a fixed productivity rate. The productivity rate is expressed in terms of output per time worked. In finer detail, it is clear that crews do not have the same productivity. Some crews are more efficient than others at specific types of work. In addition, crews on large projects will become more productive over time. Claims may include documentation of the way in which the expected productivity was not achieved due to limited site access or over-inspection.

Material

The cost of material is governed by factors that are somewhat outside the control of the scheduler. The designer specifies the required quality of materials to be installed. Procurement personnel receive bids to obtain the required materials at the least price. On-site storage is defined by the location of the jobsite and associated business and population density.

Materials for projects with sufficient on-site storage are delivered to the site and assumed to be available to workers as needed. If insufficient storage is provided, the scheduler will need to determine how to address timely material delivery in the project plan. The lack of on-site storage is often a major project constraint that should be explicitly managed. An early example of staging materials for projects was documented with the construction of the Empire State building. Having no space for an on-site steel lay-down yard, steel was delivered from the mill, still warm, directly to the site.

A material-related concern for sites that have sufficient space is placing the materials so that they require the least travel time from storage to the workface. Materials should be moved around the site as little as possible, since moving materials for non-production purposes increases costs without increasing earnings.

Logistics management and material "lay-down yards" are subjects of significant research efforts within the construction management community. The development of "lean-construction" methods addresses reductions in supply-chain waste. Technologies that employ Radio Frequency Identification (RFID) tags in lay-down yards have been shown to reduce material handling and decrease the time between delivery of, and payment for, materials.

Equipment

Various types of equipment are used on construction sites. Some of this equipment should be considered in the project plan, some need not be included. Workers, and their companies, are typically required to provide the small tools needed to complete their specific features of work. This type of equipment is generally not included in the project schedule. There are some types of equipment, however, that can affect the overall sequence and duration of the project.

The most obvious example of a constraining resource, visible at large construction sites in every city across the globe, is the construction crane. The installation and placement of cranes is critical to ensuring their maximum productivity. The CPM schedule should include, at a minimum, a milestone that includes the delivery and setup of large site-wide cranes anytime such equipment is needed. Another type of equipments that would be helpful to include in a schedule, particularly in highly constrained jobsites, are work platforms such as scaffolding. Although scaffolding is temporary construction equipment that is not installed with the project, the presence of scaffolding at a particular work-face may constrain progress on activities below where workers are employed. Conveying systems, such as temporary elevators, may also constrain tall projects. If so, including related activities in the schedule may be prudent.

Limited Resource Allocation

Limited Resource Analysis is needed to assess the impact of insufficient resources. In Limited Resource Allocation, activities are moved to ensure there are sufficient resources to complete the project. In some cases, the schedule may be delayed. The term "allocation" is used in this type of analysis since it is the job of the scheduler to determine how to assign, or allocate, scarce resources. On a project, these scarce resources could be any materials, labor, or equipment that one may need to evaluate. In the examples below, a generic resource is used for simplicity.

To introduce the topic of Limited Resource Allocation, consider the example schedule shown in Fig. 7.1. There are three activities in this schedule. In this first example, it is assumed that activities "Task 1" and "Task 3" require the same resource, "Resource A." "Task 1" and "Task 3" both require one unit of

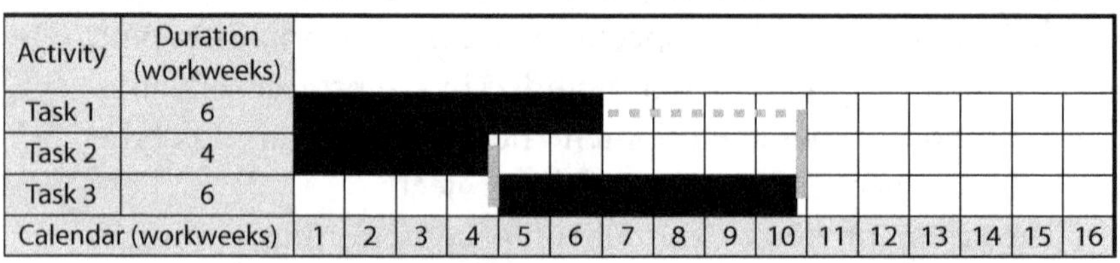

Figure 7.1 Limited Resource Analysis (Example 1).

Activity	Duration (workweeks)																
Task 1	6	1	1	1	1	1	1										
Task 2	4																
Task 3	6					1	1	1	1	1	1						
Calendar (workweeks)		1	2	3	4	5	6	7	8	9	10	11	12	13	14	15	16

FIGURE 7.2 Limited Resource Analysis with resources (Example 1).

this resource per day. To add that information to the Fenced-Bar Chart, the number of resources required per day can be provided inside the activity bar in lieu of a solid bar, as shown in Fig. 7.2.

If there is only one unit of "Resource A" available per day, then completion of the schedule must be delayed 2 workweeks. "Task 1" begins and uses the complete amount of available "Resource A" from its start through the end of week 6. "Task 2" is scheduled without difficulty since it does not require "Resource A." "Task 3" must, however, be delayed. The start of "Task 3" must be pushed back for 2 weeks from week 5 to week 7 since "Resource A" is all being used to complete "Task 1."

The procedure above assumes that once a resource is allocated to an activity, the resource will continue with an activity through its completion. Only when the allocated activity has finished will the resource be able to start the next activity. An alternative procedure, known as "Split Activities," allows resources to jump from one unfinished job to the next. In general, Splitting Activities should not be allowed due to decreased productivity of workers jumping from activity to activity.

When considering a schedule with several hundred or a thousand activities, it is not practical to use a Fenced-Bar Chart to allocate resources. For large projects computer software must be used to allocate scarce resources. The rules contained in scheduling software rank activities to determine the order in which an activity will have access to resources. If the resources are available, the activity is scheduled. If the resources are allocated to prior activities, then the next activity being considered is delayed until the completion of the higher-ranked activity. The basic role of the rule is to determine the order in which each activity should be given the resources.

Many resource allocation rules are possible. For example, giving a higher priority to activities on the Critical Path might be an effective strategy. If there are two parallel sets of activities on the Critical Path, then an additional subsorting rule will also have to be added. An example here may be that the activity with the earliest Early Start Date should go first, or that the activity with the latest Late Finish Date should go first. Usually a third rule, based on Activity ID or random selection, is also added in the case of a tie between activities on the first two criteria. The examples below illustrate the results obtained from very simple heuristics and demonstrate why the scheduler cannot rely solely on the results of such algorithms.

To investigate differences in the Limited Resource Allocation heuristics, a second example schedule shown in Fig. 7.3 adds one new activity, "Task 4," to the schedule shown in Fig. 7.2. Like "Task 1" and "Task 3," "Task 4" also requires one unit of "Resource A" per day to complete.

To conduct the Limited Resource Analysis, each activity is listed and then assigned a priority value based on the associated heuristic. Table 7.1 lists each of the activities from Fig. 7.3 and applies two different heuristics to determine the order in which each activity should be allocated "Resource A." Notice that

Activity	Duration (workweeks)																
Task 1	6	1	1	1	1	1	1										
Task 2	4																
Task 3	6					1	1	1	1	1	1						
Task 4	3					1	1	1									
Calendar (workweeks)		1	2	3	4	5	6	7	8	9	10	11	12	13	14	15	16

FIGURE 7.3 Limited Resource Analysis (Example 2).

Activity	Early Start	Total Float	Duration	ES-TF Sort	ES-EF Sort
Task 1	1	4	6	1	1
Task 3	5	0	6	2	3
Task 4	5	3	3	3	2

TABLE 7.1 Limited Resource Allocation Heuristics Sort

"Task 2" is not included in this table since "Task 2" does not use "Resource A." The fifth column of Table 7.1 lists the Early Start, Total Float Sort. This sort prioritizes activities so that those for which Early Start comes first are selected at the top of the list. If there is a tie, as in the case of "Task 3" and "Task 4," then the activity with the least Total Float is selected.

The result of allocating access to "Resource A," based on the ES-TF Sort is illustrated in Fig. 7.4. Notice in Fig. 7.4 that the ES of "Task 3" is delayed until week 7. After the completion of "Task 3," "Task 4" is pushed back until week 13.

The last column in Table 7.1 provides a slightly different sort order based on the Early Start and Early Finish sort. As before, activities with the earliest Early Start dates are selected to go first. When there is a tie, as in the case of "Task 3" and "Task 4," then the second sort, minimum Early Finish, determines which activity should be selected next. Figure 7.5 illustrates that "Task 4" begins in week 7 and that "Task 3" begins in week 10.

Industrial engineers would identify many other types of heuristics that could be applied to determine the resource

Activity	Duration (workweeks)																
Task 1	6	1	1	1	1	1	1										
Task 2	4																
Task 3	6							1	1	1	1	1	1				
Task 4	3													1	1	1	
Calendar (workweeks)		1	2	3	4	5	6	7	8	9	10	11	12	13	14	15	16

Figure 7.4 Limited Resource Allocation w/ES-TF Sort (Example 2).

Activity	Duration (workweeks)																
Task 1	6	1	1	1	1	1	1										
Task 2	4																
Task 3	6										1	1	1	1	1	1	
Task 4	3							1	1	1							
Calendar (workweeks)		1	2	3	4	5	6	7	8	9	10	11	12	13	14	15	16

Figure 7.5 Limited Resource Allocation w/ES-EF Sort (Example 2).

allocation sort order. Examples include, but are not limited to, minimum Early Start and minimum Original Duration, Early Start and minimum Resource Utilization, Early Start and maximum Resource Utilization, Total Float, and Early Start.

The application of different resource allocation rules will result in many solutions. These differences can be attributed to the specific resource limitation, layout of the schedule, and the computer sorting rule used. Given these differences, a schedule should never take the resource allocation result from a computer program at face value. The effective scheduler applies several different allocation algorithms and independently evaluates the results provided.

Unlimited Resource Leveling

Unlike the Limited Resource Allocation problem, the Unlimited Resource Leveling assumes that there are sufficient resources available to complete all tasks. Since there are sufficient resources, the project completion date may not change during the Unlimited Resource Leveling process. The goal of the Unlimited Resource Leveling algorithm is to reduce the cost of resources by reducing turnover and idle time. The two resource techniques are, however, similar in one critical way. Schedulers cannot rely solely on computer software to provide solutions. The judgment and experience of the scheduler are critical to both methods.

The example shown in Fig. 7.6 will be used to illustrate the need for and approaches to Unlimited Resource Leveling. In this example, there are three pacing activities: "Task 1," Task 2," and "Task 3." There are also three following activities, with durations

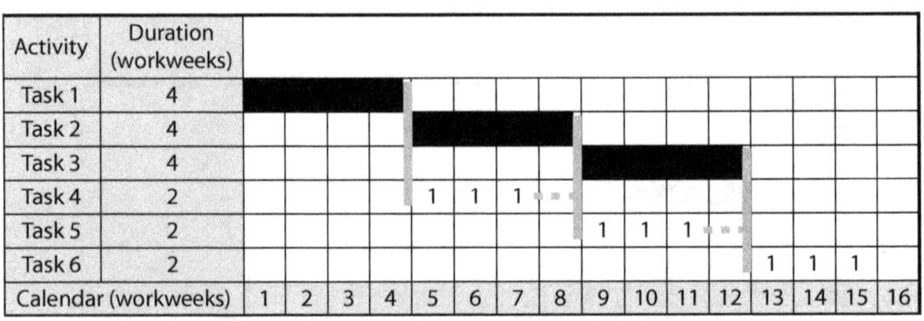

FIGURE 7.6 Unlimited Resource Leveling.

less than that of the pacing activities. "Task 4," "Task 5," and "Task 6" each require a single resource per day.

A new type of graph introduced to support Unlimited Resource Leveling is called the resource histogram. The resource histogram is developed with an *x*-axis set to the project timeline and a *y*-axis equal to the total number of resources needed each day of the timeline. Figure 7.7 provides the resource histogram derived from Fig. 7.6. As can be seen from the histogram, the resource is first needed on week 5 and is no longer needed after week 15. Within that time period are 2 weeks, week 8, and week 12, when the resource is not required.

In the Unlimited Resource Leveling, process gaps and uneven height of the histogram are of concern to the scheduler. In Fig. 7.7, one resource is needed starting on week 5, then the scheduler will need to find something for the resource to do on weeks 8 and 12. If the resource is equipment, then nonuse of the equipment will increase the price of that equipment to the company. If the resource is labor, then the crew's productivity will drop. This is because the crew will get to the start of week 7 and then decrease productivity so that they are not idled, or moved to another job, for week 8. The gap in the resource histogram for week 12 will produce another decrease in productivity. If the gap is large enough, then the resource could be moved off the job, to other work. Once off the job, the manager of the idled resource is unlikely to get the resource to return to the job for fear of additional idle time.

Uneven resource levels may also affect project quality. This is because crews may attempt to begin work in the next work area or perform work out of sequence. In Fig. 7.7, for example, if the resource performing "Task 4" were to maintain productivity, then they would have a 1-week idle time. The foreman of this crew may want to keep working on the project and begin "Task 5" prior to the completion of "Task 2." While

Resources Required	2																
	1					1	1	1		1	1	1		1	1	1	
Calendar (workweeks)		1	2	3	4	5	6	7	8	9	10	11	12	13	14	15	16

Figure 7.7 Unlimited Resource Leveling histogram.

out-of-sequence is possible, as noted in the previous chapter, such progress may require that once "Task 3" begins, the crew performing "Task 3" may need to tear-out some of the work completed in "Task 5." Anticipating such problems and finding ways to resolve them is the job of an experienced scheduler.

Scheduling software includes algorithms for Resource Leveling; however, these algorithms will generally not be successful on realistically sized projects because the algorithms rely on sorting methods that may not perform as expected. The scheduler expects that the algorithms are able to find the "best" answer, while the heuristics used can only do one simple thing—sort and schedule activities. Any computerized resource algorithm should be thoroughly reviewed and understood before it is used by the scheduler.

In many cases, the most efficient way to perform resource leveling is to start with activities defined by work type, responsibility, and work area. The resulting schedule reports allow the scheduler to directly select activities of a specific type and look at that subset of the schedule across multiple work areas. Such reports will show the gaps between the finish dates of prior activities and the starting dates of the following activities. The scheduler will learn to visualize such gaps in terms of resource histograms.

When gaps and inconsistent numbers of resources are identified in the project plan, the scheduler should suggest alternatives to the current project plan. The objective of these alternatives is to eliminate these gaps and "even-out" the planned resource use. Table 7.2 lists approaches that may be used to adjust resource levels.

A commonly used method is to delay an activity's actual start within its float range. This decrease idle time without changing the total duration of the project. In Fig. 7.6, if "Task 4" starts on its Late Start Date of week 6, then there will be no gap in the updated resource histogram. "Task 4" will end on the end of week 8 to be directly followed by "Task 5" starting on week 9. Of course, this approach may not resolve all gaps in the resource histogram. In Fig. 7.6, the scheduler can only use this technique

Technique	Possible Result
Delay activity start	May eliminate gap between repetitive tasks
Reduce crew size	Decreasing productivity may reduce gaps
Overtime use	Short term productivity increase, decrease high volume activity
Change methods	Different equipment and/or methods, may increase productivity
Subcontracting	To gain additional resources, in the case of good local resource market

TABLE 7.2 Approaches to Adjust Schedules to Level Resources

to eliminate one of the two resource histogram gaps. Either crews will be idle during week 12, stretch the work to appear busy, or leave the job to earn somewhere else. Rather than just delay an activity within its float range, delaying the start of "Task 4" by the entire gap in the subpath would resolve the issue. Such an approach was described related to Line-of-Balance scheduling illustrated in Figures 2.7 and 2.8.

Another way to decrease idle time is to slightly reduce the resource size; this will increase the duration of each activities in a series. Consider if the resource performing "Task 4," "Task 5," and "Task 6" could be adjusted so that they productivity rate would be slightly reduced. If this were the case, then these activities could be designed to take 4 weeks to complete. Such a change would eliminate both gaps in the histogram.

Reducing the duration of a longer set of activities preceding the activities with idle time is another alternative. In Fig. 7.6, additional resources could reduce the duration of "Task 2" and "Task 3" by a week. If this was possible, then the gaps between "Task 4" and "Task 5" and between "Task 5" and "Task 6" would be eliminated. The use of overtime or alternative construction methods may, for example, allow "Task 2" and "Task 3" to be accelerated.

The final alternative for smoothing Resource Leveling histograms is to simply assign the problem to someone else. Subcontracting part of the work could supplement contractor resources or allow these resources to be used more efficiently on another project.

Time-Cost Tradeoff

When the scheduler evaluates alternatives for accelerating specific activities, several options may be available. The easiest option, for example, is that existing workers add overtime hours. Next, additional workers could be added. It may even be possible that the contractor select an alternative, more expensive, construction method and completely change the original set of resources.

The scheduler will want to evaluate resourcing alternatives in several different situations. As noted in the previous section, the scheduler may want to adjust specific activities to smooth a resource histogram. The scheduler may also be faced with a request to accelerate the project. If this is the case, then the duration of activities on the Critical Path may have to be adjusted.

Liquidated Damages are contract-specified cost per day incurred by the contractor if the project falls behind schedule. If the project has been delayed, then the scheduler can help decide if Liquidated Damages should be paid, or if the contractor should pay to speed up activities that are behind schedule.

Regardless of the reason for accelerating the schedule, two types of costs will affect the analysis. The first is the cost of accelerating the schedule. This can be expressed in the cost per work period for the activity. Since the duration of the project will also be decreased as a result of accelerating the schedule, some costs that would have been incurred will be saved. These costs are the daily costs of maintaining supervision on the project. Such costs are the "Indirect Cost" of running the project. Liquidated Damages are a type of Indirect Cost. The "Time-Cost Tradeoff" technique considers both the additional costs of acceleration and reduced costs of overheads.

The objective of the Time-Cost Tradeoff is to determine which activities to accelerate, or "Crash," to produce the maximum overall benefit. There are two parts to the analysis. The first is to determine the increased direct cost incurred as the project duration is shortened. The second is to determine the savings achieved as the result of reductions to daily indirect costs. Use the following procedures to conduct the Time-Cost Tradeoff.

Step 1. Identify Normal and Crash Durations

To begin the Time-Cost Tradeoff, the scheduler must identify the normal duration and cost of each activity to "Crash." In addition, the cost of the activity, if additional resources are applied, needs to be estimated. Table 7.3 provides the example used to illustrate the Time-Cost Tradeoff. The first two columns of Table 7.3 provide the typical list of Activity and Prior activity. The next two columns provide the Normal Duration and normal cost of each task with the typical allocation of resources. For example, Task 1 is normally expected to last 3 days. Its expected cost is $2000. Summing the cost of each activity, at the Normal Duration, results in a total cost of the project as $11,000 for the Normal schedule. The contractor has the option of Crashing Task 1 to reduce its duration from 3 to 1 day. If "Task 1" is completed in a single day, the last column of Table 7.3 shows that the cost of the task will increase to $10,000.

In Table 7.3, the scheduler can see that there are many different combinations of activities that could be Crashed. The number of combinations increases when the scheduler considers that they could fully Crash some activities or only partially Crash others. Given this very large set of possible options, the Time-Cost Tradeoff procedure is a commonly applied method to determine the most cost-effective set of activities to Crash.

		Normal		Crash	
Activity	**Prior**	**Workdays**	**Cost ($US)**	**Workdays**	**Cost ($US)**
Task 1	None	3	2000	1	10,000
Task 2	Task 1	7	4000	3	8000
Task 3	Task 1	4	2000	2	10,000
Task 4	Task 3	5	3000	2	9000

Table 7.3 Normal and Crash Durations

Step 2. Calculate Cost Slope

In the Time-Cost Tradeoff analysis, is it assumed that there is a linear relationship between the number of days an activity is Crashed and increased cost of the activity. This relationship could be graphed for each activity. The duration is on the x-axis. The cost is shown on the y-axis. The slope of that line identifies the extra per day cost added to the task for each day the activity is Crashed. For example, the "Cost Slope" of "Task 1" is equal to the difference between Crash Cost and Normal Cost ($10,000 to $2000) of $8000 divided by the difference in duration of 2 days. The interpretation of this number is that for every day that "Task 1" is Crashed, there will be a $4000 increase in the cost of "Task 1." The cost of "Task 1," if Crashed 1 day will be the original cost of "Task 1," $2000, plus 1 day of $4000, for a value of $6000. Crashing "Task 1" and additional day will result in an additional cost of $4000 bringing the Crashed cost of "Task 1" to $10,000. The Cost Slope for the schedule in Table 7.3 is provided in Table 7.4.

Step 3. Prepare Fenced-Bar Chart

The next step in the Time-Cost Tradeoff process is to prepare a baseline Fenced-Bar Chart for the schedule in Table 7.4. Shading and labeling are added to the standard Fenced-Bar Chart to support the Time-Cost Tradeoff process. Each activity bar is shaded to show the minimum Crash duration and the normal duration. The baseline Fenced-Bar Chart is shown in Fig. 7.8.

		Normal		Crash		Cost Slope
Activity	**Prior**	**Workdays**	**Cost ($US)**	**Workdays**	**Cost ($US)**	**($US/Day)**
Task 1	None	3	2000	1	10,000	4000
Task 2	Task 1	7	4000	3	8000	1000
Task 3	Task 1	4	2000	2	10,000	4000
Task 4	Task 3	5	3000	2	9000	2000

TABLE 7.4 Cost Slope

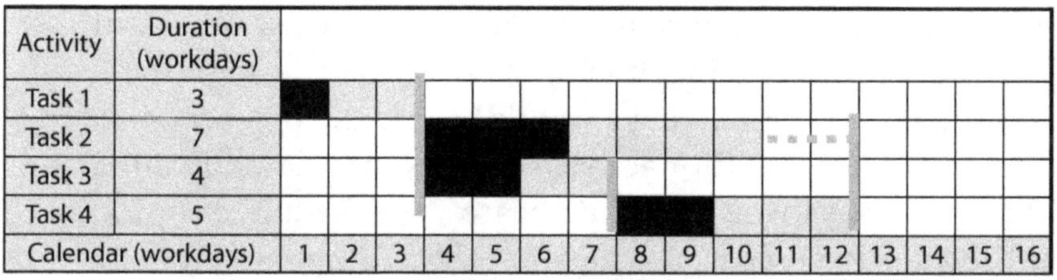

FIGURE 7.8 Time-Cost Tradeoff Fenced-Bar Chart—baseline.

One way to look at the process of Time-Cost Tradeoff is to think about activities in a schedule as if they were springs. For a given activity to move from the Normal to the Crash duration, some pressure must be applied. The pressure is higher cost. The pressure, or additional cost, applied begins as slowly as possible and increases. There is, however, one complication, that complication is that Crashing a single activity with the least Cost Slope may not result in changing the overall duration of the project. The decision of which activity or activities to Crash depends on the combination of activities on the Critical Path. Only by Crashing an activity or activities on the Critical Path will the duration of the overall project be compressed.

Step 4. Iteratively Crash the Critical Path

Begin Crashing activities starting with the least additional cost per day and continue adding costs, step by step, until no more compression is possible. This is an iterative process. As the schedule is compressed in each step, the additional cost to decrease the schedule further must be recalculated. Figure 7.9 shows the Cost-Slope Diagram for the Baseline

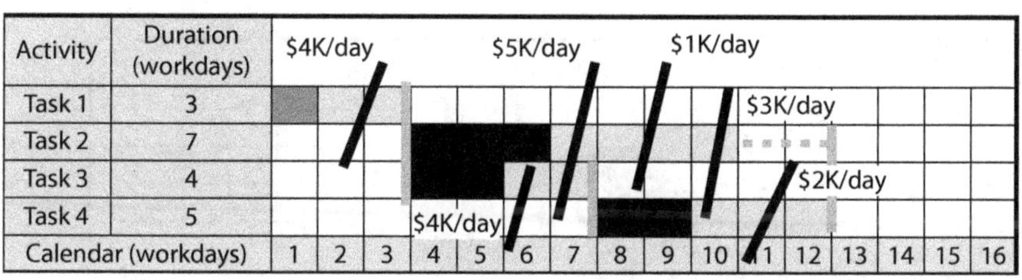

FIGURE 7.9 Cost-Slope Diagram—baseline.

schedule. On this diagram, the sum of the Cost Slopes of each activity occurring during a given day is identified. Crashing "Task 1" 2 days would decrease the schedule duration at the additional cost of $4000 per day. Crashing "Task 4" 2 days would also decrease the schedule duration at the additional cost of $2000 per day. Although "Task 2" has the least Cost Slope, Crashing "Task 2" would result in additional cost to the project but no decrease in schedule duration. Only the activity or activities on the Critical Path are Crashed at each step.

The minimum pressure that can be applied to the Baseline and decrease the overall duration of the project is to Crash "Task 4" 2 days, at a cost of $2000 per day. Completing this first step will decrease the overall duration of the project to 10 days, increase the cost of the project by $4000, and change the critical path. Figure 7.10 shows the new schedule.

The next least costly schedule acceleration is to Crash "Task 2" and "Task 4" for 1 additional day. The cost slope of this day is $3000 per day. The result of completing this step is shown in Fig. 7.11.

After the second iteration, there are only two options. To Crash "Task 1" for 2 days at a cost of $4000 per day or to

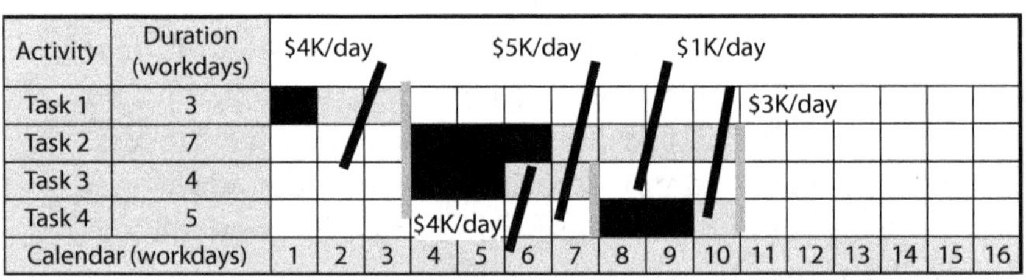

Figure 7.10 Cost-Slope—first iteration.

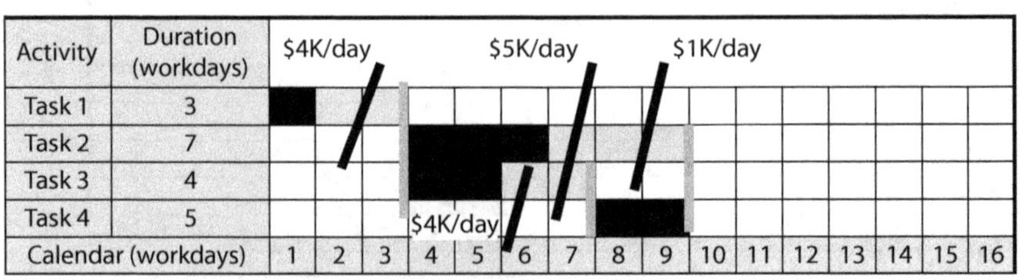

Figure 7.11 Cost-Slope—second iteration.

Crash "Task 2" and "Task 3" at a cost of $5000 per day. Both "Task 2" and "Task 3" must be Crashed together since Crashing only one would not decrease the length of the Critical Path. In the third iteration, Crashing, "Task 1" is selected as the least costly option. The result is illustrated in Fig. 7.12.

At this point in the Time-Cost Tradeoff, only two activities, "Task 2" and "Task 3" are left to be Crashed. In Fig. 7.12, the values for the Cost Slopes of "Task 2" and "Task 3" have been carried over from the original Baseline schedule. Given that the Duration and Critical Paths of the schedule have changed, the Cost-Slopes here must also be re-evaluated. In Fig. 7.12, we can see that both "Task 2" and "Task 3" are now on the Critical Path. It is also possible to compress both activities 2 days at a cost of $5000 per day, as shown in Fig. 7.13. Once this is accomplished no further schedule compression is possible.

To summarize the Time-Cost Tradeoff process it is useful to create a table documenting each step. Table 7.5 identifies the activities Crashed and the impact on the overall duration and cost of the project. The Baseline schedule took 12 days at a cost of $11,000. At the end of the Time-Cost Tradeoff process, the

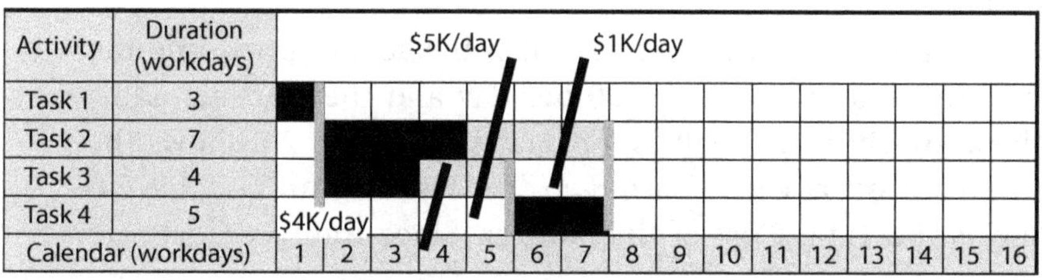

FIGURE 7.12 Cost-Slope—third iteration.

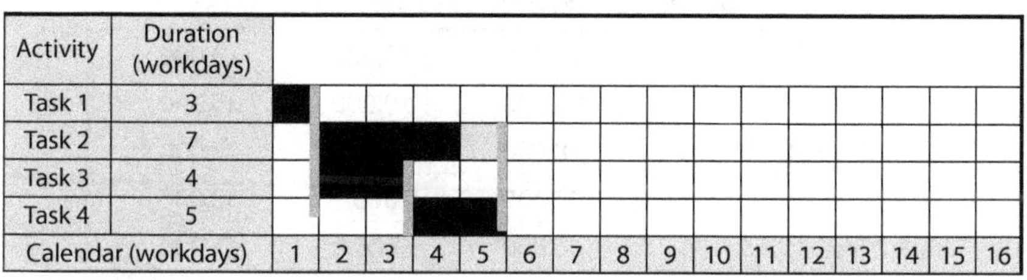

FIGURE 7.13 Cost-Slope—final iteration.

Crash Step	Activities to Crash	Crashed Duration	Project Duration	Cost Change ($US)	Direct Cost ($US)
Baseline	None	None	12	None	11,000
First	Task 4	2 days	10	+4000	15,000
Second	Task 2 and 4	1 day	9	+3000	18,000
Third	Task 1	2 days	7	+8000	26,000
Fourth	Task 2 and 3	2 days	5	+10,000	36,000

TABLE 7.5 Iterative Crashing Summary

project could be completed in less than half of that time, but at an increased cost of $36,000.

Step 5. Time-Cost Tradeoff

While project Direct Costs increase as the schedule is Crashed, Indirect Costs will decrease as the duration of the project is shortened. Some costs, such as penalty costs or incentive costs may also come into play, depending on the specifics of each project. In this step of the process, the "tradeoff" between increasing Direct Cost and decreasing Total Cost is considered. Projects with high indirect costs, penalties, or incentives make the use of the Time-Cost Tradeoff analysis a valuable scheduling tool.

First, consider the case in which the cost of providing the field office overhead is $1000 per day and there are no additional overheads, penalties, or incentives. Table 7.6 shows the Indirect Cost at each Crash Step. The last column in Table 7.6 shows the Total Cost of this project. Since Indirect Costs are

Crash Step	Activities to Crash	Project Duration	Direct Cost ($US)	Indirect Cost ($US)	Total Cost ($US)
Baseline	None	12	11,000	12,000	23,000
First	Task 4	10	15,000	10,000	25,000
Second	Tasks 2 and 4	9	18,000	9000	27,000
Third	Task 1	7	26,000	7000	33,000
Fourth	Task 2 and 3	5	36,000	5000	41,000

TABLE 7.6 Low Indirect Cost Summary

low, the information provided in Table 7.6 shows that there will be no financial benefit for Crashing the schedule.

Next, consider the impact of increasing indirect costs and owner penalties. Assume that the additional cost of jobsite supervision is $2000 per day (Indirect Costs) and an owner penalty of $3000 per day is assessed for each day after day 8. These changes are reflected in Table 7.7 in the updated Indirect Cost, Damages, and Total Cost columns. Notice that the minimum Total Cost appears on day 9. One lesson to take away from the Time-Cost Tradeoff is that there may be value to the contractor in paying more for specific activities: to pay less overall. The lesson for owners is that adding penalties to a project may be insufficient to motivate an on-time project completion if the cost of the penalty is less than the cost of accelerating the work.

Like the S-curve, the Time-Cost Tradeoff has a representative shape. Figure 7.14 provides the graph of Table 7.7. The *x*-axis of the Time-Cost Tradeoff graph is the total project duration. The *y*-axis in Fig. 7.14 is cost in dollars. Reflecting the four columns in Table 7.7, there are four lines sown in Fig. 7.14. The line closest to the *x*-axis is the line that shows the total damages accrued by the project for each day. Moving upward in the *y*-direction (at x = 8), the next line charts the project's indirect costs. The line that is second from the top is the direct cost. As the project duration increases direct costs decrease and indirect costs and damages increase. The "U-shape" of the total cost line is representative of projects that complete a Time-Cost Tradeoff analysis. The least costly overall project is found at the total cost curve's minimum.

Crash Step	Activities to Crash	Project Duration	Direct Cost ($US)	Indirect Cost ($US)	Penalties ($US)	Total Cost ($US)
Baseline	None	12	11,000	24,000	12,000	47,000
First	Task 4	10	15,000	20,000	6000	41,000
Second	Tasks 2 and 4	9	18,000	18,000	3000	39,000
Third	Task 1	7	26,000	14,000	–	40,000
Fourth	Task 2 and 3	5	36,000	10,000	–	46,000

TABLE 7.7 High Indirect Cost Summary

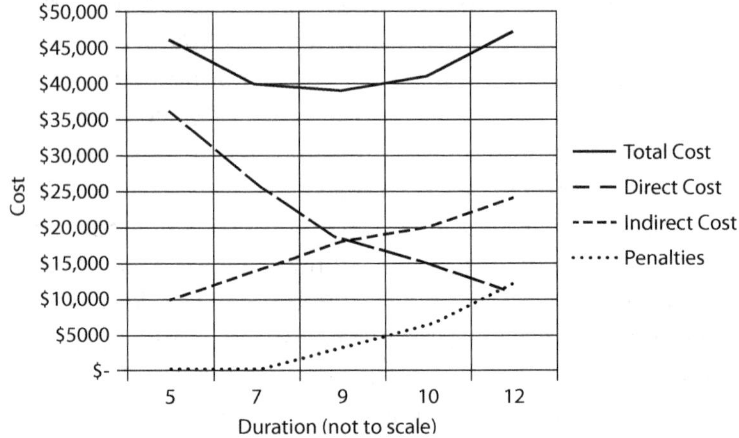

Figure 7.14 Time-Cost Tradeoff graph.

Cash-Flow Analysis

The preceding section introduced how to include the contractor's direct and indirect costs into scheduling practice. In this section, a technique for an expanded analysis of cost is introduced. The typical use of Earned Value Management on fixed-price construction contracts is helpful to predicting incoming cash flows; however, it does not model cash outflows. If the construction project is to be financed, either internally or externally, then a Cash-Flow Analysis should be conducted to model both income and outflow.

There is a significant delay between when a contractor incurs a cost and when that cost is reimbursed. Contractors typically pay their workers weekly, or biweekly. Materials, tools, and equipment must also be purchased or rented. Contractors accumulate these costs over a period of time and submit their invoice for payment to the owner's representative. Depending upon the owner and or local law, the owner will typically have 30 days to process that payment. As a result of this payment delay, contractors must know how much money will be needed at what point in the project to ensure that workers and suppliers are paid on a timely basis.

Cash-Flow Analysis assists in predicting, based on the current status of the project, the expected working capital needed to fund the project—the expected break-even point for the project.

Cash-Flow Analysis may also assist in the evaluation of working capital and break-even point under varying sets of subcontracting options, supplier payment terms, and interest rates. The steps required to perform the Cash-Flow Analysis are described below.

Step 1. Estimate Material, Labor, and Equipment Cost

While the expected costs of a project will have been defined in a winning bid, the application of these costs (and not simply the Earned Value) for the work performed must be directly linked to schedule activities. The following rules-of-thumb can assist the project team to allocate contractor, subcontractor, and material on-site costs.

1. When defining activities, separate activities that combine prime-contractor and subcontractor work. This is required because there is a time lag between the prime-contractor payment period and the subcontractor payment period.
2. Separate activities that combine equipment or material that will be stored in an off-site warehouse or stored on-site prior to installation. This is required since the prime-contractor should request payment for the delivery of the equipment and material.
3. Any construction activity that is greater than the payment (30 calendar or 20 working days) should be split into multiple activities. Making activity durations less than the typical monthly payment invoice cycle will force a more accurate evaluation of the status of an activity.
4. The exception to splitting work activities into multiple activities of less than 30 calendar or 20 working days is for activities that, by contract, incorporate unit cost schedules.
5. Any activity that includes more than one unit cost item, or bid item, should be split until each activity has only one unit cost item. This is important since it will be easier to reconcile unit cost items if they are clearly delineated as separate activities.
6. The delivery and installation of specialized construction equipment, such as cranes, jobsite trailers, etc.

should be identified as separate schedule activities since owners may allow payment for the lease of this equipment independent of the cost of installation activities.

7. The cost of jobsite items for which costs are paid up-front may also be removed from monthly site overhead costs and included as individual activities supporting the mobilization to the site. While a "mobilization" activity is typically not riembursed by owners items such as site fencing and signage that are specifically required in the contract to be installed at the start of the project, may, for some owners, be paid directly.

Step 2. Allocate Direct Cost

In a basic schedule, the scheduler will include the Earned Value with each activity in the schedule. In the Cash-Flow Analysis, the scheduler assigns both the Earned Value and the direct costs of each activity to the schedule. The following rules-of-thumb will be helpful for this process.

1. Create a value-loaded, time-scaled schedule in a Bar Chart or Fenced-Bar Chart. Note that the location of bars within their possible range of float should reflect the way that one may plan to do the work. Use of the default early-start sort for laying out the Bar Chart may be unrealistic and result in increased cost of project working capital.
2. The major vertical grid lines of the schedule should be in 30 calendar or 20 working days to reflect the payment period.
3. Split the costs of activities that span a given payment cycle into their prorated portions based on a dollar per day estimate of activity Earned Value.
4. Sum activities' costs that one may plan to complete during the each payment cycle. Transfer this information to a spreadsheet to speed up the remaining steps in the Cash-Flow Analysis procedure.

Step 3. Allocate Overhead Costs

Overhead costs must be paid in order to perform the work, but are not directly tied to one specific activity. There are two kinds of overheads: jobsite and home-office overheads.

Jobsite Overheads

Calculate the cost of small tools, gang boxes, permits, telephone services, site Internet connections, security guards, and site superintendent for the whole project. These are examples of some of the site-specific overheads. The project may also include other costs such as trailers, fencing, and signage. Depending on the way that the items are paid for, these may be either individual activities or monthly charges. Once the total cost of jobsite overheads is calculated, divide that total by the number of payment cycles in the project. If payment cycles are in months, as they typically are, then the cost per month to run that project is determined.

Home-Office Overheads

There are also costs that must be paid related to the functioning of the home-office. The largest of these costs is likely to be a portion salaries of home-office staff. Company equipment used across all projects should also be included as part of home-office overheads. Additional home-office overhead costs include utilities, legal and accounting fees, computer and software costs, company vehicles, travel, etc. The allocation of home-office costs is more complex than allocating jobsite overheads. This is because the home-office may be working on several projects at the same time. The standard way of allocating home-office overhead is called the "Eichleay" formula. This formula proportionally allocates cost of the home-office overhead based on the size of the project compared with the total volume of work at the company. The formula itself is [(Total Home-Office Cost)/(Number of Project Payment Cycles)] * [(Current Project Cost)/(Total Cost of All Projects)].

Step 4. Add Markup

Contingency costs are those costs that could be incurred given the uncertainty of the project. While a detailed risk analysis could be conducted, experienced company owners use percentages to estimate the "known unknown" additional costs for the project. Contingency is often based on factors such as project type, project location, design specificity, market conditions, and experience with the owner. Stated another way, contingency is the contractor's compensation for the uncertainty in the contract requirements: the higher the uncertainty, the higher the contingency, and therefore the higher the price of the contract. The lower the contractor's risk in accepting the project, the lower the price of the contract. Issues directly under the owner's (or owner representative's) control that may lower construction costs are the specificity of design requirements and clear understanding of working relationships and administrative requirements imposed by the contract.

For international builders, the largest contingency costs can result from the potential mitigation of delays or due to social unrest or political instability. At a minimum, this contingency is found in the bid price as on-site security. In very remote locations, or difficult working environments, contractors may also need to provide complete worker encampments with water treatment, medical, and dining facilities.

Another kind of contingency costs, particularly for projects in uncertain economic conditions, is inflation. Projects that require the purchase of materials from multiple countries also have a currency risk that can be estimated. The costs and overheads identified in the previous steps are typically calculated based on the current value of a single currency and under expectation of zero inflation. An adjustment for expected future inflation should be added to projects being built in uncertain fiscal settings.

Profit is the return on the investment needed by those who providing provide funding for the project. While different industries have different average levels of profit, there is a bottom line that often helps students understand the idea of profit. Consider a business that has $10,000 to invest in something. If they put that money in an investment account, they may be able to realize a 5% annual return and receive 10,500 at the end of a year. If they put that money in a construction contract, then the return they receive should be at least as much as that received by the investment account. The riskier the construction project, the higher rate of return, since if the risker project fails, then funds may be lost.

Step 5. Calculate Expected Earned Value

At each month add the markup to the previously calculated cost and overhead figure. This sum will be what the owner should pay for the work accomplished. Since the schedule is used to drive these costs, the Earned Value per month can be developed for the entire duration of the project. The representative shape of the cumulative, expected Earned Value, is called the S-curve.

Step 6. Determine Payment Offset

Assuming that the billing is accomplished the first day of the month after the work is completed, the owner's processing of that check will take some time. A good owner will pay within 30 days of receipt of the invoice. A poor owner (and one for whom contractors increase their contingency costs) may or may not provide prompt payments.

Step 7. Calculate Overdraft

At the start of the project there is no payment for Earned Value because the construction has not yet been completed. During this period, the contractor must still pay salaries and suppliers. The difference between the total of the project earned and the total the owner has paid is the "overdraft."

Step 8. Calculate Interest Payment

The organization that pays for the cost of the work before the owner has paid for the value of the work result is essentially loaning money to the project team. The project team will not get this money for free. The company providing the loan would evaluate the loan the same way that the contractor evaluates any other possible investment. The way that the company providing the loan makes a profit on that loan is to charge "interest." Interest accrues during the project as long as there is an overdraft on the project.

Step 9. Calculate Retainage

During the project, the owner may hold back a portion of the payment as "retainage." This amount of value earned but not paid will often be held in the case of open punch-list items or other deficiencies in the quality of the installed work. Some owners also hold retainage to ensure that once the contractor breaks-even on the job, the remaining tasks at the end of the job are completed.

Step 10. Determine Break-Even Point

The steps above can be used to develop the expected outflows and inflows of funds throughout the life of the project. If the contractor is operating a typical contract, then costs accrue before being paid for earnings. Eventually, however, earnings catch up with and exceed the direct costs of the project. The difference between the costs and payment is the profit. Most contractors working fixed-price contracts do not receive more earnings than payments until the final payment On such projects, the contractor will not show a profit on a project until the final pay invoice.

CHAPTER 8

Planning a Mid-Rise Building

The previous chapters of this book introduced the arithmetic of CPM, its basic techniques, and rules-of-thumb needed to create and use schedules effectively. Behind these tools and techniques was the idea that a schedule is a simple model of a construction project with a focus on sequence and duration. The novice scheduler can perform the arithmetic calculations of CPM, but may not be able to create a good model of the underling construction project. The experienced scheduler can think about the way that the building will be constructed, identify project-specific constraints that affect planned construction methods, and translate this information into a CPM schedule model. The difference between the novice and experienced scheduler is, in part, being able to identify those aspects of a project that could affect sequence and duration.

In this chapter, a simple project will be developed to help new schedulers begin to see how to apply the techniques presented in this book. The example building described in this chapter is a three-story, mixed-use building in a small town located in the Midwestern United States. In thinking about this example as it unfolds, the reader should consider what types of issues are and are not discussed. The creation of an effective schedule is as much about what is left out and approximated, as it is about the specific aspects of each project that are included.

Examine the Contract

Regardless of the project, the work to be completed is defined by the contract. The contract defines both the quantity of the work to be performed, typically found on the design drawings, and the quality of the components and work results, typically found in the construction specifications. For traditional design-bid-build projects, the plans and specifications are used to prepare the cost estimate for the project and from that construction bids are prepared. The bidder winning the construction award then begins the process of building the project. One of the first steps in this process will be to secure the services of a construction scheduler.

At this first stage of the scheduling process, the design drawings will be the most important part of the contract documents for the scheduler to review. The objective of this first review is to identify the vertical and horizontal decomposition of the project. By understanding the overall structure of the building, the scheduler starts the work by visualizing crews coming on the project and moving through each work area in turn as the next workspace becomes available.

The location of the project is one of the aspects of the contract that should be identified at this early stage. Location is a critical factor in the development of a project schedule. Consider the potential impact of location on a project built in a remote Rocky Mountain town. Access to building materials may be delayed due to the remote location. The mile-high elevation will also have an impact on the productivity of both man and machine.

Understanding the impact of seasonal weather will be important to ensure that the building can be enclosed prior to inclement weather. In some cases, inclement weather will result from cold temperatures; in other cases, heat and humidity may cause problems.

In addition to weather, access to basic infrastructure is also of critical concern for scheduling construction projects. Consider the differences in delivery times for basic construction materials at a site within 5 miles of the United States interstate highway system with a site that has no paved access roads and is 200 miles from an interstate highway. Even the most basic resources, such as water, may be in question if the project is to be constructed in a desert region.

For this example, it is assumed that the building is being built in a small city in the mid-west United States. As a result, the scheduler can assume there will be no difficulty in obtaining construction materials. Materials and workers will be able to get to the jobsite using the existing road networks.

List Major Contractual Obligations

The two most obvious contractual obligations to be included in a schedule are the construction Notice to Proceed (NTP) date and the construction duration. The construction duration and NTP define the date before which the contract must be complete, the Contract Completion Date. The NTP and the Contract Completion Date provide the total time bounds of the project and should be directly included as activities in the schedule. Use a Start-No-Earlier-Than milestone on a zero day duration "NTP" activity as the first activity in the schedule. The Start-No-Earlier-Than milestone will ensure that the project schedule does not begin until the contractually allowed date. If the scheduler is developing the schedule after NTP, then at this stage, the scheduler should ensure that the initial schedule calculation date (Time-Now) is set to the contractual NTP date.

The last activity in the schedule should have a Finish-No-Later-Than milestone on a zero day duration activity called "Complete Project." The Finish-No-Later-Than milestone will, once the schedule has been calculated, let the project team know if the project is ahead or behind schedule based on the Total Float value of the Critical Path. If the minimum Total Float in the schedule is negative, the project will take longer than the project completion milestone date. This means that the project is behind schedule. If the minimum Total Float in the schedule is zero, it means that the project will take exactly as long as the time available to complete the project. If the minimum Total Float is positive, then the project is expected to be complete before the contractually required completion date. In other words, if the Total Float is positive, then the project is ahead of schedule.

The duration of both of these milestone activities or any others should be set to zero days. The zero duration for these milestone activities ensures that the project can work the first full day and the last full day of the contract. If non-zero duration milestones are included, there could be a week or more tied-up in these "informational" activities.

In addition to the "NTP" and "Complete Project" activities, milestones should be added to correspond with all interim contractually required dates. In mixed-use, mid-rise buildings, for example, there may be contractual requirements that requires first floor retail spaces to open to the public as soon as possible, with office and residential floors opening after completing the remainder of the building. In our example, it is assumed that there are no interim project milestones.

Large projects are often split into contractually defined phases. Renovation projects may have a phase where a building addition is constructed, followed by a move-in phase, with a third phase being the renovation of the existing (now vacated) spaces. A zero-day duration activity should be included for the start and end of each phase. If there are contractually required start and ending times for the phases, then the start and end activities will need to have fixed milestones.

Many projects include the construction of multiple buildings. Such contracts will identify any requirements for the required completion of individual buildings. These interim milestones should be included as zero-day duration activities with fixed Start-No-Earlier than and Finish-No-Later than milestones as needed to reflect the specifics of the contractual obligations noted in the contract.

A type of contractual obligation found in many construction contracts requires the contractor to provide a number of administrative submittals prior to the start of construction. Often such requirements are found in the General Provisions of the construction contract. These submittals including quality control plans, safety plans, erosion protection plans,

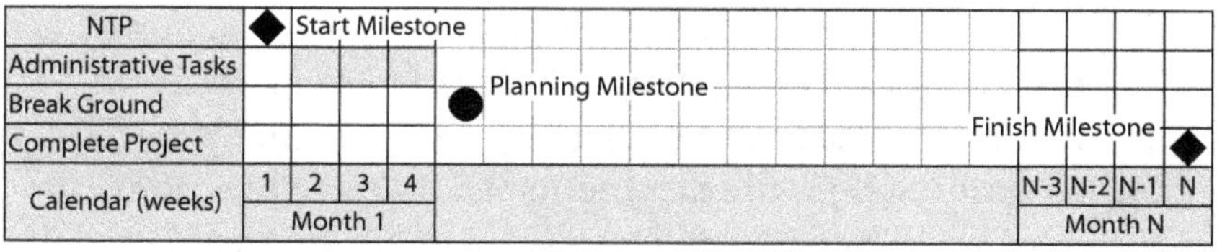

Figure 8.1 Working schedule with project milestones.

site access plans, etc. must be approved prior to groundbreaking. Along with these administrative submittals, other activities where cost would be allocated to project overheads can be directly included in the project. Examples of these project mobilization costs include: payment for construction performance bond, site trailers, site telecommunications, fencing, and signage. Including contractually required mobilization activities with appropriate cost loading in the progress schedule may allow the contractor to recoup some of the up-front costs needed to get a project underway. Policies differ on payment for mobilization activities from owner to owner; therefore, the scheduler should coordinate with the contractor and owner's representative to determine which of these mobilization activities may be cost loaded. Figure 8.1 provides an initial template into which most building projects, including the example building project described in this chapter, could fit.

Divide Project into Major Work Areas

Within the context of a given building or project phase, the contractor and the scheduler should identify the major physical work areas through which crews will need to traverse from the start to the end of the project. For the example mid-rise building, there are three floors above ground and no below ground floors. The example building has a single roof area that includes mechanical equipment. The major work areas for the example building will be the foundation, each of the three floors, and exterior closure (including the roof). Depending on the interest of the project team in communicating their plan with other project team members, adding starting and ending planning milestones with zero-day durations for each of these major work areas could be included in the schedule. Fixing the Early Start and Late Finish dates for the starting and ending activities using milestones should not be done unless there are contractually binding dates included in the construction contract.

The example project is a single mid-rise building. If there were multiple buildings or phases, then each building or phase should be considered by the scheduler as separate work area, even if the contract did not specify start and end dates for each building or phase. A standard, recognizable, name for these starting and ending activities should be consistently used across each of the buildings or phases.

Hammock activities may also be included to quickly report summary work area schedules to the owner. Hammocks take the early and late dates of their linked prior and following activity. This allows Hammocks to reflect planned and actual progress. Figure 8.2 extends the generic schedule shown in Fig. 8.1 by adding Hammock activities for each of the example building's floors.

Hammock activities linking the start and end of major features of work such as foundation, floors, and closure allow the overall progress of the schedule to be summarized in a Bar Chart – hiding the complexity of the underlying CPM schedule. At this stage of the project, the durations of the activities are not known; therefore, time units are not listed on the *x*-axis of the Figure 8.2

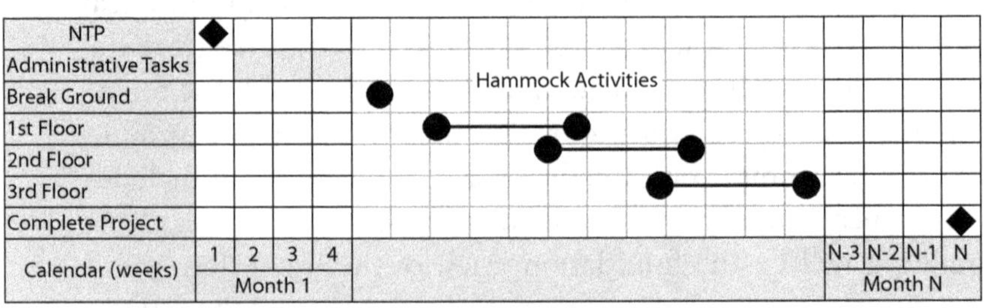

Figure 8.2 Working schedule with work area Hammocks.

Building Method Implies Sequence

In the previous step, the overall work areas through which crews will proceed as they complete the project were defined based on the physical constraints of the building. At this stage, conversation between the scheduler and contractor establish the precise way that the work is to be accomplished. The focus of these discussions should be on the method that the contractor is planning to use to construct the building. Even when the structural system of a building is specified by design, the contractor may be able to build the project in several different ways. It is the method selected that defines the required crews, and thus defines the rate of production for each crew.

Often a contractor will identify the crew that is to perform the job and then talk about how the crew will perform the work. It is a subtle distinction, but one worth noting that a crew could do the work in several different ways but that it is the method used which drives the sequence and duration of activities within and between different work areas.

Typically, the most important method decision documented by the scheduler is the contractor's selected technology to build the structure. For the example building discussed in this chapter, the structure is specified to be cast-in-place concrete with exterior masonry. The contractor's method for our example includes placing concrete for the foundation and then using a small crane to transport concrete to floors above the first floor using buckets. Another contractor may have selected to pump the concrete. In other locations, a contractor may have chosen to use a continuous stream of human-powered wheel barrows to place the concrete.

Based on the design of the building and the selected construction method, a typical sequence for construction above the foundation can be developed both for the structure and for the structure and activities within a given floor. Table 8.1 identifies one possible set of activities need to complete a given floor of the example building.

Following decisions about how the structure is to be built, the scheduler will need to discuss how other crews will access the site and how their materials will be moved up the building.

Activity
Place Concrete Formwork
Place Rebar and Concrete
Finish Concrete
Cure Concrete
Strip Concrete Formwork
Install Metal Studs
Rough-In Plumbing
Rough-In Electrical/Communication
Install Drywall
Finish Drywall
Paint Walls
Install Carpet
Install Cabinets
Install Appliances
Finish Plumbing
Finish Electrical

TABLE 8.1 Template of Activities Per Floor

Management of the material lay-down yards is essential to an efficient construction project since materials that are moved several times require additional labor costs and increase the percentage of materials that are damaged and must be replaced. For small mid-rise buildings, such as the example project, it will be assumed that a combination of external lifts for the lower floors and to the stairwell will be used to move building materials vertically within the building. A closed trash shoot will also be used to ensure that transportation of waste will not interfere with building materials moving up the stairs.

During these discussions, the scheduler should also compare decisions made by the contractor with assumptions made by the contractor's estimating department during bidding. If the contractor's estimator selected a construction technology that differs from that planned, additional work will be needed to determine any expected deviation from the cost estimate and construction schedule. The art of scheduling is about understanding differences between the planned or expected project and the actual project being constructed. Anticipating differences in cost estimates and schedules between bidding phase and construction phase is often the first opportunity for schedulers to practice their art.

At this stage, the schedule that has been developing through paper sketches and communicated to other team members via email attachments may be entered into a computerized CPM program. There are many such programs from which to choose. Most project schedulers are constrained in their selection of which products to use due to the dominance of a few widely used scheduling software programs. The purpose of the CPM Tutor is not to provide training on the use of these tools, but to educate the scheduler to decide whether the software in question has the features needed to efficiently and effectively create an effective CPM schedule.

As the generic patterns of activities are being included in the schedule for the technology selected, the scheduler should ensure that the activities are coded to allow specific activity subsets to be selected and reviewed without having to sift-through the overall schedule. This coding should be accomplished through the use of both activity identification number and additional activity code fields. Schedulers will have to consult their software users' manual to determine how to create and maintain these codes.

For our example project, activity numbers will be set to range between 001 and 999. The scheduler will use the placement of information in these numbers as the basis to identify where the activity takes place and the type of work being completed in that activity. The hundred's positions will be used to designate the major vertical sections of the building. The ten's and one's places will identify the type of work being performed. Numbers can also be reserved for pre construction and post construction activities. Numbers should be skipped to ensure that if new activities need to be added, they can be added without renumbering the existing standard set of activities. The first column of Table 8.2 adds generic activity numbers for each of the standard activities in the schedule. As noted in the column heading, the numbers listed are the numbers of the

Number 0XX's	Activity	Crew Code
05	Place Concrete Formwork	Formwork
10	Place Rebar and Concrete	Concrete
15	Finish Concrete	Concrete
20	Cure Concrete	Concrete
25	Strip Concrete Formwork	Formwork
30	Install Metal Studs	Carpenter
35	Rough-In Plumbing	Plumber
40	Rough-In Electrical/Communication	Electrician
45	Install Drywall	Carpenter
50	Finish Drywall	Finisher
55	Paint Walls	Painter
60	Install Doors	Carpenter
65	Install Flooring	Flooring
70	Install Cabinets	Carpenter
75	Install Appliances	Appliance
80	Finish Plumbing	Plumber
85	Finish Electrical	Electrician

Table 8.2 Template Activity Numbering and Crew Code

activities that will appear in the ten's and one's positions in the final activity numbers.

Another type of information vital for the project team to know is the set of activities to be accomplished by different crew types. While some computerized scheduling systems support heuristic resource scheduling, it is possible to gain much of the value of those algorithms simply by adding crew type or Responsibility Codes. The third column of Table 8.2 shows the activity coding for crew type.

The identification of what type of crew to use is also contingent upon the status of the project as union or non-union project. Care should also be taken to identify crews, even if that work is to be subcontracted. This is because the subcontractor's schedule will be just as dependent upon access to work areas and crew production rates as the prime contractor's crews. A full evaluation of the schedule at this initial stage of the project is needed, even if the ultimate responsibility for the work will fall to someone other than the prime contractor.

As the project progresses, there may be crews that perform more than one type of work. Keeping the work type as part of the activity identification number, and the crew type as a separate activity code, is important for larger projects to evaluate crews' flow through the project. In this small example project the Electrical Crew can to identify that its activities comprise the following list: 140, 185, 240, 285, 340, and 385. This is because these are the activity numbers for the "Electrician's Rough-In," X40, and "Finish Electrical," X85, activities on the first, second, and third floor of the example building.

The scheduler should also ensure that activities that will constrain progress are included in the list of activities. Notice the "Cure Concrete" activity in Table 8.2. This activity requires no crew and has no value; however, failure to include this activity will result in significant errors in the overall construction schedule. This is because the formwork on the floor below cannot be removed during the curing period. Failure to heed such physical constraints is a recipe for failure. The premature removal of forms has been responsible for several building collapses.

Given the list of example activities and floors, it is now possible to quickly build the entire main-sequence production schedule by adding the hundred's places to those activity templates. For example, the second floor "Finish Plumbing" activity would be identified as activity 280. The first floor "Install Drywall" activity would be identified as activity 145. As noted in the number column heading, the numbers shown in Table 8.3 reflect the number that will be added to the hundreds' place in the final activity number.

There are, of course, many other activities needed in a project that take place prior to the start of work on the first floor. Notice that in Table 8.2, there is also no allowance for preconstruction activities, foundations, site work, work on the roof, or project completion activities. To accommodate these types of activities, the per-floor activity-numbering scheme can be

Number X00's	Activity's Work Area
0	Mobilization
0	Foundation
1	First floor
2	Second floor
3	Third floor
4	Roof
5	Site
6	Closeout

TABLE 8.3 Completed Activity-Numbering Scheme

extended to include the numbering conventions for work areas outside the vertical structure of the building.

While standard types of areas in the building and administrative activities can be identified by using the hundreds' position in the schedule, there are also specific areas of the building the scheduler may want to consider separately given the complexity of that work or the impact of that work on the overall project. Two such examples in a typical three-story building are "Install Elevators" and "Install Low-Slope Roof." Work on the structure of the elevators is completed, in the case of our example, through the carpentry activities already identified in Table 8.2; however, the procurement of the elevator system is a long lead item that will require special consideration. A preconstruction activity related to procure hydraulic elevators will, therefore, be added to the schedule. Per the activity-numbering convention used in this example, the number for this activity should be less than one hundred.

Another long-lead equipment procurement item, the roof-mounted chiller, will be added to the schedule as activity "Install Rooftop Units." The addition of this equipment is important since a crane will be required to be on-site when the equipment is placed on the roof. Many projects have been delayed due to the need for specialized equipment. The scheduler should consider all activities that require cranes or special lifts as these activities are likely to cause unexpected delays.

Another special consideration is that of activities supporting the exterior closure of the building. For this example project, reinforced masonry with brick facing is specified as the exterior surface of the building. Additional activities related to "Exterior Wall" and "Exterior Windows" will be added to the schedule. These activities will take place within each of the three floors, but also should be specifically identified as separate managed tasks using an activity code because the completion of this type of activity must precede completion of many of the subsequent activities within a floor. For example, interior finishing work should not be attempted if the weather is too cold or too wet. Table 8.4 adds two activities for exterior closure on each floor, Activities 27 and 28. Notice that the gaps in the original activity numbering (Table 8.2) allowed these two new activities to be easily added to the schedule.

Depending on the interests of the scheduler, and the extent to which the construction season may limit outside work, the scheduler may want to add a planning milestone for the completion of exterior closure including the completion of the roof. This milestone can simply be a zero-duration activity, called '"Complete Exterior Closure" that will move with the schedule to reflect the expected building closure.

The last step needed to round-out the per floor activity template is to identify the sequence among activities. Table 8.5 identifies the required sequence of the activities. In this example project, only the conventional Finish-to-Start schedule was used. In general, the scheduler should avoid more complex sequences since they take more time to set up and will be less well understood by the team.

Number OXX's	Activity	Crew
27	Exterior Walls	Masonry
28	Install Exterior Windows	Carpenter

TABLE 8.4 Additional Exterior Closure Activities

Number 0XX's	Activity	Prior Number
05	Place Concrete Formwork	–
10	Place Rebar and Concrete	05
15	Finish Concrete	10
20	Cure Concrete	15
25	Strip Concrete Formwork	20
27	Exterior Walls	15
28	Install Exterior Windows	27
30	Install Metal Studs	25
35	Rough-In Plumbing	30
40	Rough-In Electrical/Com.	30
45	Install Drywall	35, 40
50	Finish Drywall	45
55	Paint Walls	50
60	Install Doors	55
65	Install Flooring	55
70	Install Cabinets	65
75	Install Appliances	70
80	Finish Plumbing	75
85	Finish Electrical	75

Table 8.5 Per-Floor Activity Template with Priors

When adding each template to the build the overall schedule additional, the sequence to model the physical and crew constraints between floors also needs to be added. While the Place Concrete Formwork is shown, in Table 8.5, without any priors in the per floor template, this is not true in the context of the overall project. Each floor starts by placing the concrete pan and associated formwork. On the first floor, this work is preceded by the foundation. On floors two and three, as well as the roof, that level cannot begin until the formwork on the previous floor has been removed. While the per-floor activity list need not show such sequence, the completed CPM schedule must include it or the physical constraints of construction would not be reflected in the schedule's calculations.

Additional sequence should be added between activities to reflect how crews proceed through the project as each floor is created. For example, if the crew to perform the Install Metal Studs activity is to start at the first floor of the building and move through each of the other two floors without leaving the job, then the Install Metal Stud activities or any other such activity will also have to be tied to their preceding activity on the floor below.

It may also be noted that different contractors may want to run the projects in slightly different ways. A major motivator in such decisions will be considerations of the types of rework that could be required on projects. Drywall specialty contractors may be interested in installing one face of the stud walls prior to Rough-In Electrical and Rough-In Plumbing. Depending on the placement of appliances, it may also reduce touch up work if the doors are installed after, instead of before, the appliances. These are the types of decisions that the scheduler cannot make without direct consultation with prime and subcontractors.

Given the cost and time spent on Heating, Ventilating, and Air Conditioning (HVAC) equipment on most projects, an experienced reader may also be interested in the location of this equipment. For this small, residential building it is assumed that a package unit will be provided for each unit. This package unit can be assumed to be included in the plumbing and electrical activities. The single rooftop condensing unit is added to our example schedule. If such an assumption is not correct, or if there is a specialty subcontractor for this task, an additional activity(ies) and sequence should be included.

The scheduler should also be aware that buildings with mixed may have different durations and possibly different sequences that require slight changes to the per floor template for a given work area. In this example building, all floors are considered to be residential; if the first floor were, however, to be commercial space to be leased and fit-up later for businesses such as restaurants, banks, or small retail shops, the activities on the first floor may be quite different. Even in this example, a lobby with mailroom, security/maintenance office, and a small seating area would require some modification of the per-floor activity template for the first floor.

Find Productivity, Duration, Cost

Once decisions have been made by the contractor about the type of technology to be used, the number of available crews and information about building material management have been resolved, the scheduler can estimate crew productivity and begin to identify the duration for each of the activities. Following the estimation of duration, the scheduler will develop an "earned value schedule."

The estimates of schedule duration made by the scheduler are often accomplished independently of the contractor's construction estimate; however, this is not the best situation as the estimator may have superior knowledge of the project that should be consulted. The ideal situation would be for the scheduler to develop assumptions about crew productivity from an independent assessment of how the project should be run and then discuss the assumptions with the estimator. Developing scheduling assumptions independently allows the scheduler to direct questions to the project team in areas where the scheduler's assumptions do not match the contractor's bid estimates. At a minimum, the scheduler should have access to the contractor's bid estimate to see if there were notes or other assumptions documented in the construction estimate.

Table 8.6 extends the per-floor activity templates presented earlier and augments them with an example of the type of information a scheduler may provide following a detailed analysis of the plans, specifications and assumptions about expected technologies, required crew sizing, and resulting crew productivity. Durations are identified in workweeks. The expected owner's earned value is expressed in United States Dollars. These durations and should not be taken as representative of actual construction project values but provided as an illustration.

Now that the sequence and duration of a generic set of activities for each of the three floors of the building has been developed,

Number OXX's	Activity	Prior Number	Duration (Workweeks)	Value ($US)
05	Place Concrete Formwork	–	1	5000
10	Place Rebar and Concrete	05	1	40,000
15	Finish Concrete	10	1	2000
20	Cure Concrete	15	3	0
25	Strip Concrete Formwork	20	1	5000
27	Exterior Walls	15	2	50,000
28	Install Exterior Windows	27	1	30,000
30	Install Metal Studs	25	2	30,000
35	Rough-In Plumbing	30	1	20,000
40	Rough-In Electrical/ Communication	30	1	30,000
45	Install Drywall	35, 40	2	40,000
50	Finish Drywall	45	1	5000
55	Paint Walls	50	2	10,000
60	Install Doors	55	1	25,000
65	Install Flooring	55	1	75,000
70	Install Cabinets	65	1	50,000
75	Install Appliances	70	1	100,000
80	Finish Plumbing	75	1	30,000
85	Finish Electrical	75	1	50,000

TABLE 8.6 Per-Floor Activity Template with Typical Duration and Value

a Fenced-Bar Chart for these template activities is shown in Figure 8.3. Like the other Fenced-Bar Chart figures in this book, Fig. 8.3 was developed using a spreadsheet program. This figure communicates to the project team each of the activities that needs to be accomplished, in order, to complete the project as early as possible.

The per-floor template shown in Figure 8.3 has two paths. The first path is the top set of activities going from the upper left of the figure to the lower right. The first path plans the progress of crews as work on one of the floors but excludes the exterior closure activities, 27 and 28. The second path includes only the exterior closure activities, 27 and 28.

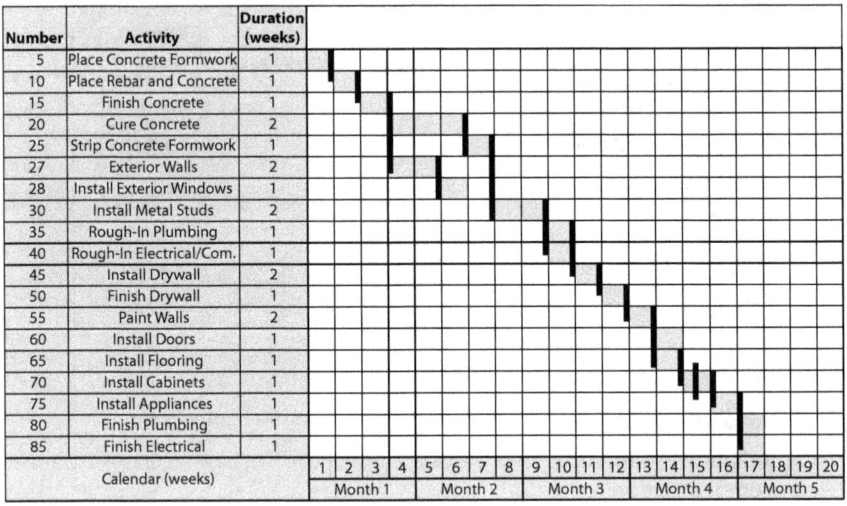

FIGURE 8.3 Per-floor activity template Fenced-Bar Chart.

Notice that not all the necessary logic of construction sequence that would need to be in a final schedule is provided in this Fenced-Bar Chart. An example of such missing sequence is the logic that would require the "Exterior Closure" of the building to precede the "Install Drywall" on the building's interior partition. Such a constraint might be critical for rapid completion of a project that must be enclosed prior to the start of an inclement weather season. Other logic that would be included on a completed, computerized, schedule version of this example project would be the logic that links together crews to ensure that work a floor above cannot begin until work for that crew on the floor below has been complete. While this simple example will assume that crews move from the bottom of the building to the top, this may not always be the case.

There are several activities, noted previously in the consideration of the building plan, that are not directly part of the per-floor activity template. These activities include milestone and Hammock activities. Site, roof, and closure activities also need to be added. For our example project, Table 8.7 provides the remaining activities.

Number OXX's	Activity	Duration (Workweeks)	Value ($US)
001	NTP	ES Milestone	0
001	Insurance	0	5000
005	Procure Elevator	4	250,000
050	Break Ground	Milestone	0
060	Clear Site	2	25,000
065	Foundation	4	50,000
100	Complete First Floor	Hammock	0
200	Complete Second Floor	Hammock	0
300	Complete Third Floor	Hammock	0
400	Complete Roof	Hammock	0
405	Install Low-Slope Roof	2	150,000
410	Install Rooftop Units	1	100,000
499	Complete Exterior Closure	Milestone	0
999	Complete Contract	EF Milestone	0

TABLE 8.7 Additional Project Activities

Depending on the cost of equipment and the quality of the finishes in a building, it may be relevant for the prime contractor to procure these expensive products. The advantage of such a procurement policy is that prime contractor's cash earnings could be generated earlier on the project. The disadvantage is the additional cost for warehousing and security on delivered but not installed equipment. In this example, the cost of procurement was not "front-loaded" to the owner but directly included in each of the "installation" activities in the project.

It would be helpful to note that at this point the entire project schedule has been developed simply by making lists of needed activities and their sequence and duration. Novice schedulers begin the process of scheduling by thinking that a particular piece of scheduling software should be immediately employed. The premature use of scheduling software with full-color, professionally appearing charts and graphs will actually detract from the real interest of the scheduler in getting information from the project team about the way in which the project is to be constructed. The novice scheduler should remember that the plan comes before the schedule, and the schedule comes before its computerized implementation.

Calculate Initial Schedule

Some construction contracts require the contractor to send in a "preliminary" schedule within 30 days of NTP. The initial schedule noted here is for the overall project, and not for the contractually required, preliminary schedule. The value of creating a preliminary schedule is for the contractor to quickly identify all administrative and long-lead procurement items required before project breaks ground. Ensuring that such submittals are begun, along with the identification of any administrative activities that the owner's representative will be willing to pay, such as performance bond, or need to perform, such as authorizing site access, will help jump-start the project cash flow.

The schedule so far is shown in Fig. 8.4. This Fenced-Bar Chart is not complete as links between exterior closure activities, or the addition of temporary exterior weather barrier activities, have not been included. Also, the links between the activities between floors are not present. These links of activities between floors, if present, would help define the latest that the activities could begin without delaying the overall project, i.e., the Late Start and Late Finish dates. With that information the Total Float is calculated and from there the identification of the Critical Path.

Even without the formal calculation of the Critical Path, it is possible to make an educated guess about activities expected on the Critical Path. To conduct this inspection, look at the activities on the schedule and "slide" their start or extended duration so that their finish dates move to the right one or more workweeks. Those activities for which movement to the right of the time line will have the result of also moving following activities, ultimately delaying the overall completion date of the project, are on the Critical Path. Based on this analysis, the example project's Critical Path runs from the initial mobilization activities, through the completion of the structure of each of the floors and finishes with the completion of the third floor.

Compare the impact of the movement of first, second, and third floor structural activities "to the right" with the movement of finish activities on the first two floors. Notice that moving work on the interior partitions later in the project would not be expected to delay the overall project. Of course, such assumptions cannot be pushed too far, because the complete set of schedule logic is not present on the Fenced-Bar Chart.

A useful concept for the project scheduler is the idea of a "pacing" set of activities and a set of "nonpacing" activities. This concept was previously introduced using a Line-Of-Balance presentation in Figures 2.6 and 2.7. The productivity of those activities on the Critical Path would "set the pace" for crews whose work must fill-in behind them. The crews working on the noncritical, or nonpacing activities may slow their productivity without affecting the duration of the project so that they do not have to repeatedly move off and then back on the job.

Once all activities have been added to the schedule with duration and cost specific to the project, the initial CPM schedule can also be calculated using computerized project scheduling software. The novice scheduler may think that this step would be the first step in the scheduling process, but the experienced scheduler knows understanding the constraints identified in previous stages, as noted above, is more important than the mechanical calculation of the CPM schedule.

Once the computerized version of the schedule is created, specific schedule reports can be produced to assist in the management of the project. While specific software may have different names for such reports, reports helpful for analysis and communication of the contractor's project plan include the following content:

- *Activity Report.* Lists activities sorted by Activity Number. Important for the project team to understand the entire context of the schedule and to ensure completeness.
- *Milestone/Hammock Report.* Lists Milestone and Hammock activities. Important for communication with project stakeholders.

Number	Activity	Duration (weeks)
1	NTP	0
2	Insurance	0
5	Procure Elevator	4
50	Break Ground	0
60	Clear Site	2
65	Foundation	4
100	Complete First Floor	17
105	Place Concrete Formwork	1
110	Place Rebar and Concrete	1
115	Finish Concrete	1
120	Cure Concrete	2
125	Strip Concrete Formwork	1
127	Exterior Walls	2
128	Install Exterior Windows	1
130	Install Metal Studs	2
135	Rough-In Plumbing	1
140	Rough-In Electrical/Com.	1
145	Install Drywall	2
150	Finish Drywall	1
155	Paint Walls	2
160	Install Doors	1
165	Install Flooring	1
170	Install Cabinets	1
175	Install Appliances	1
180	Finish Plumbing	1
185	Finish Electrical	1
200	Complete Second Floor	17
205	Place Concrete Formwork	1
210	Place Rebar and Concrete	1
215	Finish Concrete	1
220	Cure Concrete	2
225	Strip Concrete Formwork	1
227	Exterior Walls	2
228	Install Exterior Windows	1
230	Install Metal Studs	2

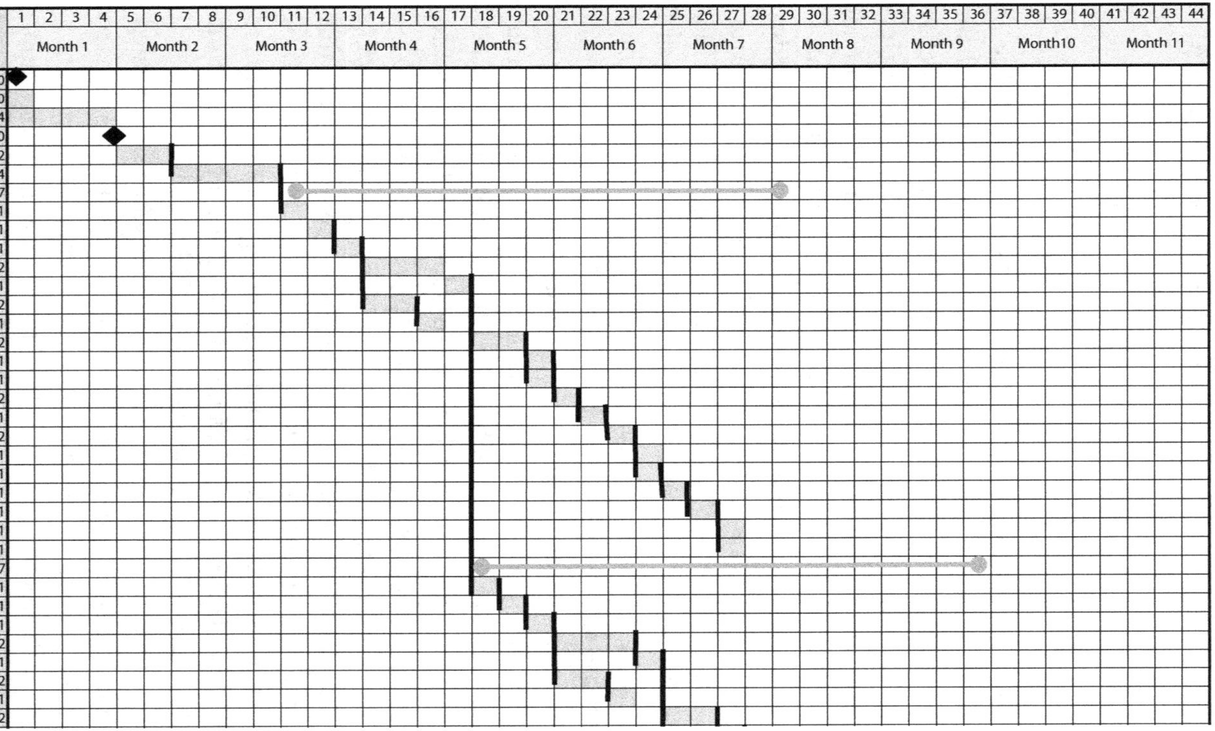

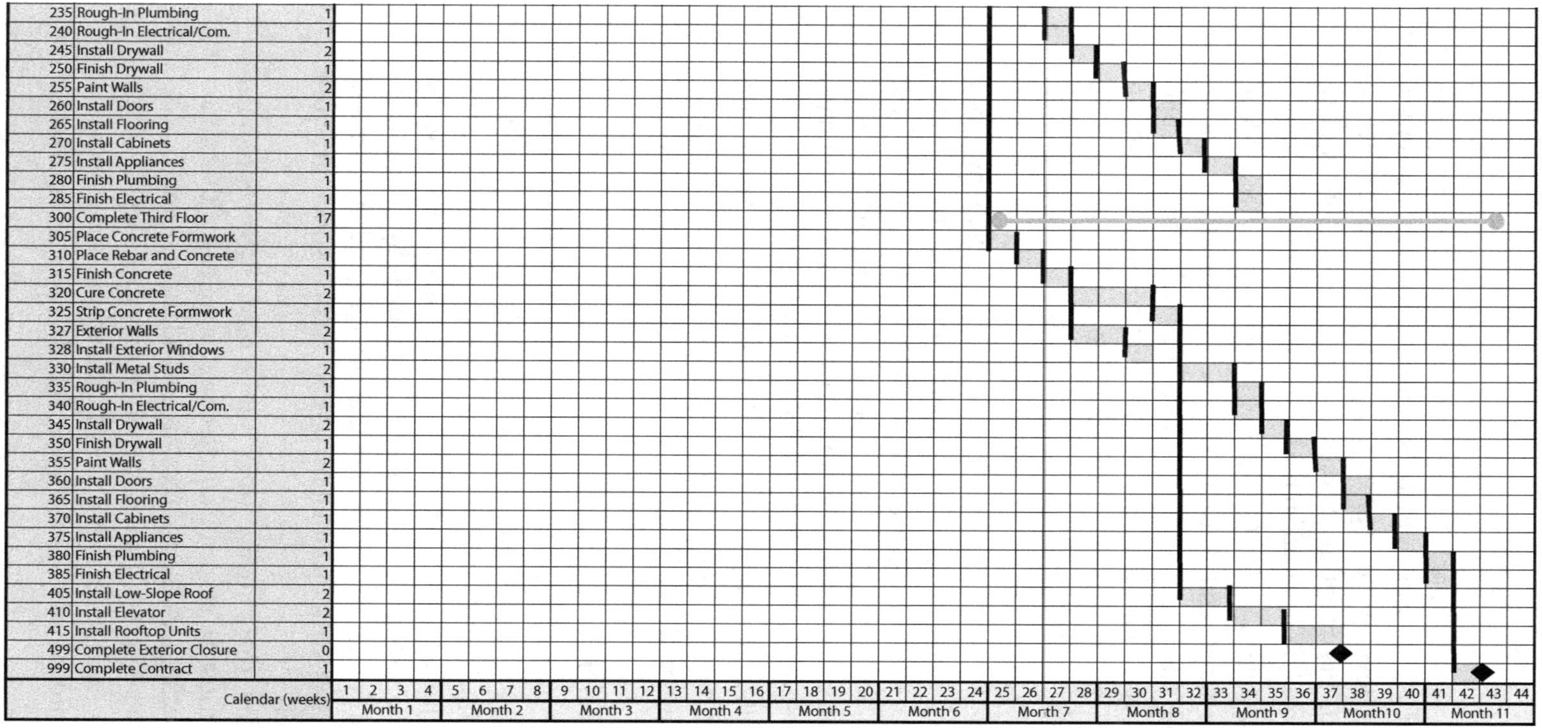

FIGURE 8.4 Example project Fenced-Bar Chart.

- *Early Start Report.* Lists activities sorted by Early Start, Total Float, and Activity Number. Important to understand how work is to proceed. This report limited to activities to begin within 2 weeks of the current date is often called a "look-ahead" report.
- *Total Float Report.* Lists activities sorted by Total Float, Early Start, and Activity number. Important to understand the most Critical Paths through the schedule. Report should be reviewed in total to ensure all paths have reasonable levels of Total Float. Following initial review of the schedule, the Total Float report may be limited to paths for which float is most critical.
- *Crew Schedule Report.* Lists activities sorted by Crew Type, Early Start, Total Float, and Activity Number. Important to communicate with crew chiefs about expected work to be performed. When report is limited to activities that begin within 2 weeks of the current date, the report is often called a "look-ahead" report.
- *Activity Logic Report.* Lists activities sorted by Activity ID. Important to confirm that proper sequence has been entered in the schedule.

Schedule Adjustments

Following the initial schedule calculation, the scheduler may need to make adjustments to the schedule. Several different types of adjustments are typically required. First, the schedule must be adjusted if contractually required milestones cannot be met with the technology and crew choices made by the contractor. The contractor should ensure that the start and durations of subcontracted work included in the contracts of subcontractors may be met.

If needed, these changes may not be made by the scheduler alone, but must be made by all affecting members of the project team. These changes and the reasons behind them should be documented in an updated schedule and in narrative form. In Table 7.2, techniques for adjusting activity durations were identified. These techniques may be used by the scheduler, with consultation by prime and subcontractors, to evaluate the required adjustments. These schedule adjustment techniques include:

- Resource Allocation and Leveling
- Time/Cost Trade-off
- Total Project Cash Flow

One shortcut that can be used to evaluate resources is the use of crew scheduling reports, along with trial-and-error schedule changes to determine whether the productivity rates of the crews can be reasonably adjusted to remove excess float in nonpacing activities.

While computerized methods may prove very helpful on larger projects, simple tools such as the spreadsheets used to develop the Bar Charts shown in this book may also assist in this analysis. To use spreadsheets for this task, the scheduler will need to have both the Early Start Fenced-Bar Chart and the Late Start Fenced-Bar Chart. Figure 8.4 provides the overall project schedule for the Early Start project. The scheduler can determine, based on implicit knowledge of the Late Finish constraints on the schedule, cut and move the activities to their Late Finish times. This results in a Fenced-Bar Chart for the Late Finish schedule Note that this Late Finish schedule does not include the complete set of schedule logic available through computerized schedule methods.

With these limited spreadsheet schedules for the Early Start and Late Start schedules the scheduler can conduct some initial resource and cost analysis. While such analysis does not precisely comply with the textbook presentation of resource and cost analysis, the methods provide a practical middle ground between failure to anticipate problems with resources and earnings, and a fully technically accurate analysis of the project using complex software that often must be reverse-engineered to understand changes to the schedule made by computer software algorithms.

For small, uncomplicated projects, the Early Start and Late Start schedule spreadsheets can be used for a rough version of Earned Value and Unlimited Resource Leveling analysis. Begin the Earned Value analysis by making a copy of the original schedule worksheet to a new worksheet. Next, enter the average

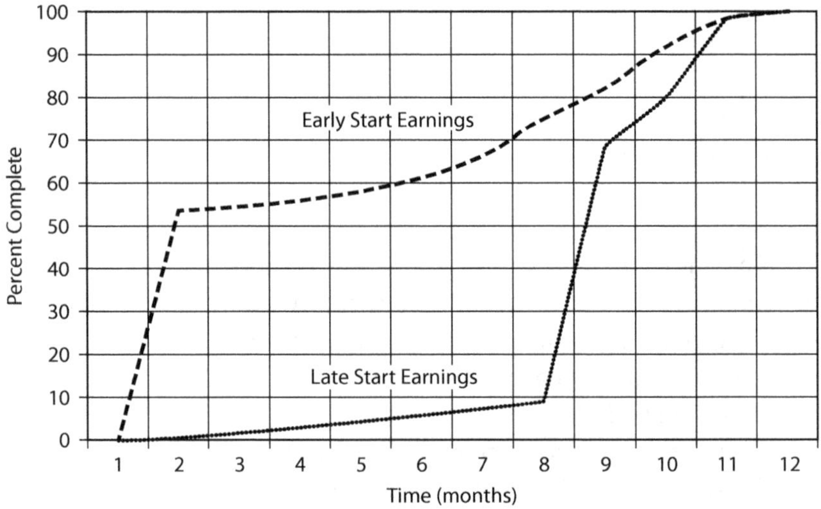

FIGURE 8.5 Earned value envelope.

weekly Earned Value in each activity's bar. Summing each week at the bottom of the sheet and then summing each week for a given month provides the S-curve that reflects Early Start Earning. Copy this Early Start Earnings sheet to a new worksheet and then cut and past the activity bars to their Late Start Dates. The summary of costs for such a schedule provides the S-curve for the Late Start Schedule. Figure 8.5 shows range of earnings that could be expected on the project. The top line in Figure 8.5 reflects estimated earnings based on the earliest possible plan. The lower line in Figure 8.5 reflects estimated earnings based on the latest possible plan.

For a simplified version of an Unlimited Resource Leveling procedure, begin by making a copy of the original schedule for the crew that one may wish to study. Next, enter the number of resources (materials, labor crews, or equipment) needed each week of that activity. Summing the information at the bottom of the sheet allows one to identify any gaps within the resource plan for that crew. An example is discussed in the following paragraphs.

On many projects, the prime contractor is responsible for Carpentry activities. The example below will, therefore, be focused on Carpentry activities. The top chart in Fig. 8.6

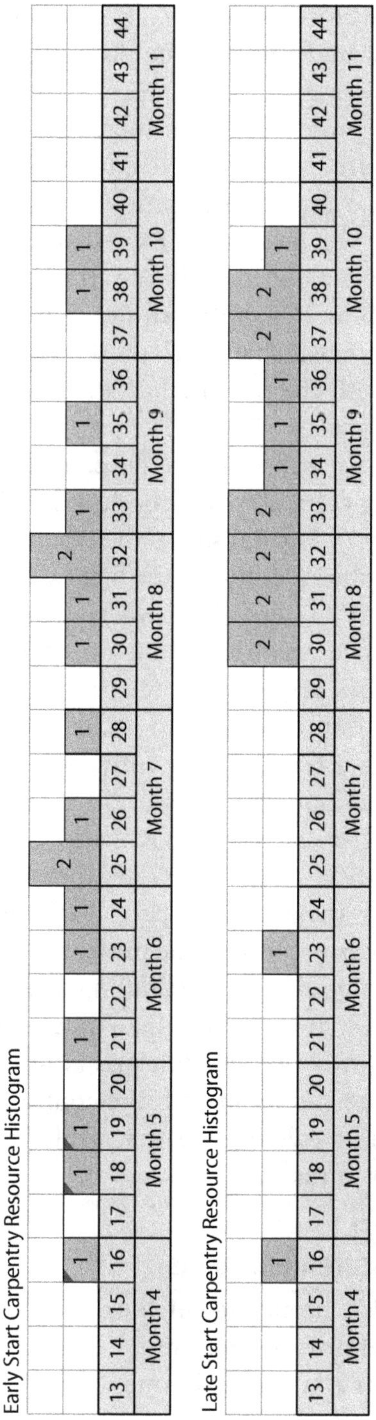

Figure 8.6 Carpentry crew histograms.

provides the Early Start histogram for Carpentry crews. Ideally, the carpentry crews would come on the site and work through the project without having any idle time. Given that crews are often able to see the progress on the jobsite and adjust productivity rates to ensure that they have enough work to do to remain employed on a project, the prudent project manager could guess that there will be a loss of 8 crew-weeks of productivity added to the project due to inefficient project scheduling. In addition, weeks 25 and 32 both require two crews on the site. If only one crew is available, work on those related activities will be delayed with the potential of affecting subsequent activities.

As an alternative, the scheduler could consider the potential impact of delaying carpentry activities until their Last Start dates. The second part of Fig. 8.6 shows the resulting carpentry crew histogram. With the exception of the first two crew requirements on week 16 and 23 needed for wall rough-in, the project manager sending two crews to the site on week 30 will only encounter a 4 crew-week productivity delay in weeks 30 to 39.

Activities cannot simply be made to occur on their Early or Late Start Dates without coordinating with others on the project. With an analysis of both the Early and Late Start schedule histogram, the scheduler can begin to make recommendations about the total number of carpentry crews that the contractor might want to have on the job and expected windows of when these crews would move on and off the project.

While there are computerized scheduling techniques available for conducting resource analysis, schedulers must be very careful to ensure that it is the scheduler making the decisions about the project, in consultation with the contractor, and not the software performing calculations that while arithmetically correct, have no basis in reality.

Include All Important Procurement Items

Adjustments may also be needed if long-lead procurement items, such as generators or elevators, cannot be reasonably procured within the allotted time. There may also be difficulties in obtaining access to key construction equipment needed to start a particular feature of work. Nonwork activities can be added to the schedule to track these key resources.

Notice that the Early Start schedule in Fig. 8.5 has significantly accelerated earnings over the earnings expected with the Late Start Schedule, one of the major reasons for the early spike in earnings for the Early Start schedule was, in fact, a procurement activity. The procurement and delivery of the elevators to the site within the first month of the project resulted in a large difference in the shape of the expected earnings curve from the traditional S-curve. The scheduler should note that on real construction projects, elevators or other critical or long-procurement items are, however, unlikely to be available at the beginning of the project.

Conclusion

Effective project schedulers know how to apply the basics of the CPM to create a good model of the construction project. They know when to group activities composed of many steps, and when to break activities into smaller parts corresponding to problem activities that must be managed individually. The example in this chapter has shown that some nonproduction activities such as contractually required milestones and procurement activities should also be included to assist in communicating the constraints on the project. In addition to communicating with the owner, noting changes in these issues as the project progresses demonstrates to the project team the scheduler's expertise in considering contingencies to cover typical problems that occur on many types of projects.

Effective schedulers also know that overloading clients and project team members with too much scheduling information will lead to ineffective communications. Schedulers can use Hammocks to communicate high-level progress with clients and tenants. For trades, the scheduler can use activity reports and schedules that show only those activities for a given subcontractor or crew and demonstrate how those activities fit within the larger project context.

As schedulers work on larger and larger projects, they should remember that team communication is the most important project management skill. The Critical Path Method, the techniques introduced in this book, and software used to prepare CPM schedules are just the tools to facilitate this communication.

CHAPTER 9
Exercises

Even when the theory of CPM is well understood, the step-by-step method needed to develop and use CPM diagrams may not be immediately clear. This chapter provides exercises and discussions that illustrate the procedure to create schedules from activity lists. Additional exercises provided without discussion may be used as homework or self-study problems. Step-by-step solutions to all exercises are available to qualifying educators.[1]

CPM Diagramming

For each of the problems in this section, draw an Arrow Diagram, Fenced-Bar Chart, and Precedence Diagram. The drawing should be neat, orderly, and clearly labeled. The minimum number of crossovers and Dummy Activities should be used. The Arrow Diagram and Precedence Diagram should be developed with a single starting and ending node or activity. The Precedence Diagram should be calculated using a 7-day workweek calendar with no holidays. The legend for the Arrow and Precedence diagram shall be that used earlier in this text.

Problem 9.1. The first step in any CPM diagramming problem is to produce a Layout Drawing. The Layout Drawing helps by finding the activity layout that reduces crossing activities, fences, and precedence arrows. It also provides a double check

[1]Send an email request containing a link to the official course syllabus showing that (a) the requestor is the instructor of the referenced course and (b) that the *CPM Tutor* is a required text for the course.

Activity	Prior Activity	Duration (Workdays)
A	–	5
B	–	3
C	A, B	2
D	A, B	4

TABLE 9.1 Activity List for Problem 9.1

that solution represents the Sequence shown in the problem. To produce the Layout Drawing, start by creating a small activity node for each of the starting activities in Table 9.1. Starting with each of those beginning activities, add the following activities. If precedence arrows cross, consider revising the order of the previous activities to reduce or eliminate crossovers. Repeat the process until all activities are shown on the Layout Drawing. The Layout Drawing for Prob. 9.1 is shown in Fig. 9.1.

Begin the production of the Arrow Diagram by adding one starting i-node for the schedule. Enter the node number 5, as that node's number. Next, draw the starting activities from that initial starting node. In this exercise, the activities that start the schedule are "Task A" and "Task B." Add these activities to the diagram and then consider where their j-nodes should be placed. Since "Task A" and "Task B" share the exact same set of followers, they must also share a j-node. Having activities with the same i-node and j-node will, however, violate a constraints found in ADM. That constraint is that every activity in a schedule has a unique i-node and j-node number pair. To resolve this situation, a Numbering Dummy must be added to either "Task A" or "Task B." The example, shown in Fig. 9.2, adds a Numbering Dummy at the end of "Task B."

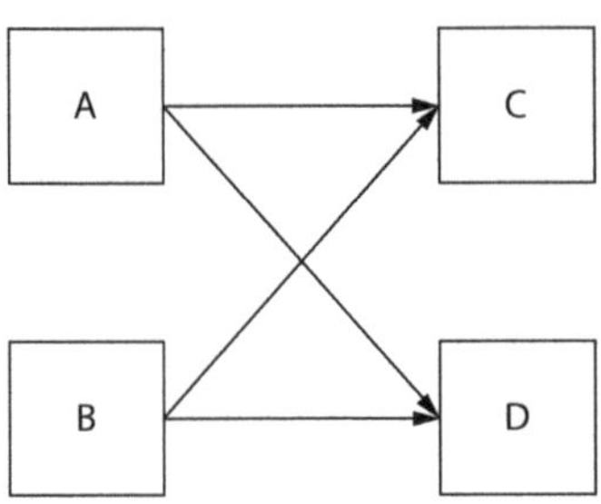

FIGURE 9.1 Layout Drawing.

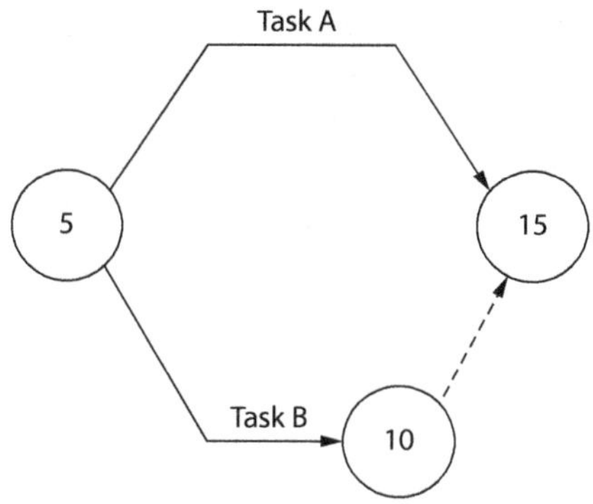

FIGURE 9.2 Added Numbering Dummy.

Once these nodes have been added, their node numbers can be set. Since best-practice dictates that node numbers should skip five digits, the j-node of "Task B' is set to 10. Since, by definition, the j-node must always be greater than an activity's i-node, the j-node at the end of "Task A" is set to 15.

Complete the Arrow Diagram layout by adding "Task C" and "Task D" to the schedule. Notice that since "Task C" and "Task D" have exactly the same set of priors and both conclude the schedule, so must end at a shared j-node. When adding a final node to the schedule one may add one more Numbering Dummy as show in Fig. 9.3.

Following the layout of the Arrow Diagram, the symbols to support CPM calculation can be added. Mirroring the ADM diagram in Figure 3.14, the activity's duration is placed below each activity arrow, centered under the name of the activity. At each node, the "Early Event Time" and "Late Event Time" boxes are placed. The Early Event Time boxes are placed above the

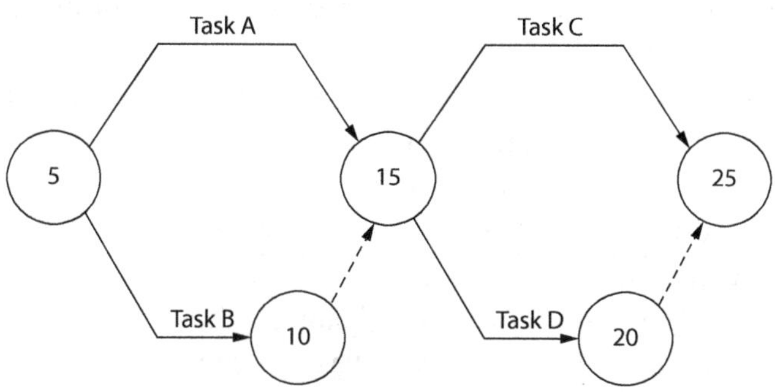

FIGURE 9.3 Complete Arrow Diagram layout.

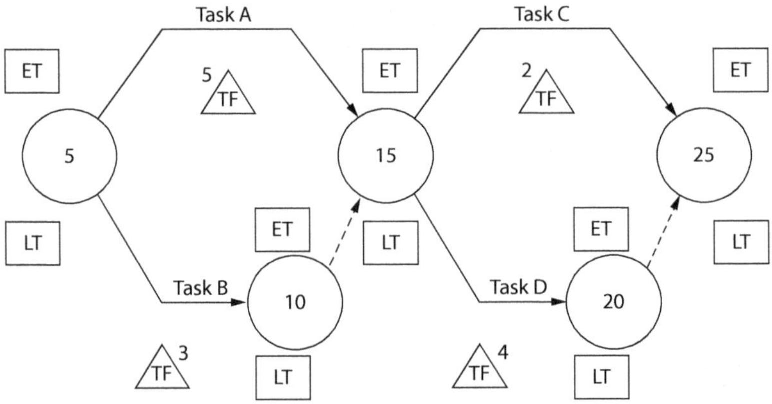

FIGURE 9.4 Arrow Diagram ready for calculation.

node. The Late Event Time boxes are placed below the node. Finally, the Total Float symbol is placed next to each activity's duration. Figure 9.4 shows the schedule once these symbols are placed on the schedule.

To simplify manually produced Arrow Diagrams, the CPM Tutor did not include the Activity Name, Duration, or Total Float of the Dummy Activities (i.e. i-node, j-node pairs 10-15 and 20-25) in Figure 9.4. This omission was made because Dummy Activities have zero day durations and their calculations are, therefore, obvious. Technically, ADM uses Dummy Activities during its calculation procedure. As a result, many course instructors may require information about Dummy activities to be explicitly shown on the diagram - just like any other activity. If instructors do not state their preference, students should explicitly include an Activity Name (i.e. "Dummy"), Activity Duration (i.e. zero), and the calculated Total Float for all ADM Dummy Activities.

At this point, the CPM schedule calculation can proceed. To begin, the Arrow Diagram's first i-node is set to an Early Event Time of zero and normal arithmetic calculations. The Forward Pass is conducted from the start of the project, at the left-hand side of the diagram. The Early Event Time at node 10 is the Early Event Time at the i-node of "Task B" plus the duration of "Task B," in other words three. The Forward Pass for the Dummy Activity following "Task B" can now be calculated. The result is that one possible Early Event Time for node 15 is three. Node 15, however, is set by the later of either the Early Event Time of the Dummy Activity or "Task A." In this case, the Early Event Time for node 15 will be five. The Forward Pass continues through until node 25.

Once the Forward Pass has completed, the Backward Pass can begin. To begin the Backward Pass, set the Late Event Time at the last node to equal the node's Early Event Time. Then, move from right-to-left through the schedule to calculate the Late Event Times. The Late Event Time at node 15 is the earliest of the possible incoming Event Times from either "Task C" or "Task D." Continue through the Backward Pass to node 5. The last step is to calculate the Total Float. The complete CPM calculation is provided in Fig. 9.5.

Next, the Precedence Diagram for Problem 1 can be developed. Begin with the Layout Drawing to guide the placement of each activity on the PDM diagram. Good scheduling practice requires that a schedule have only one starting activity and only one ending activity. In Prob. 9.1, there are two starting activities and two ending activities. As a result, either "Task A" or "Task B" should be chosen as the starting activity and Either "Task C" or "Task D" as the ending activity. A Start-To-Start sequence should be added to the beginning of the project to tie together the starting activities. A Finish-To-Finish sequence should be added to the end of the project between "Task C" and "Task D". When adding Start-to-Start or Finish-to-Finish sequence, always show the Lead and Lag Durations. Here the Lead and Lag Durations

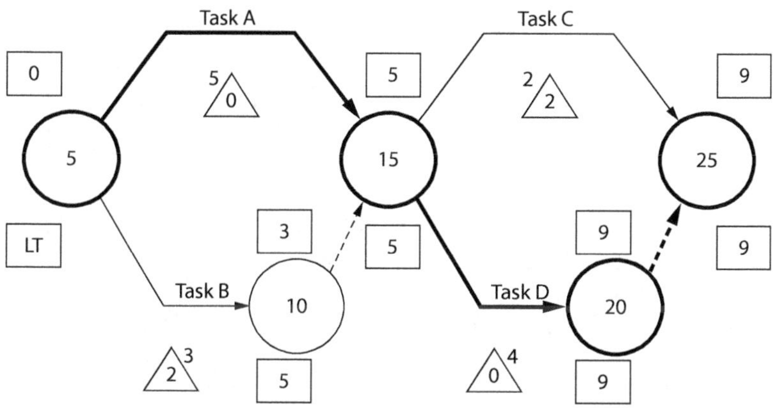

FIGURE 9.5 Complete Arrow Diagram.

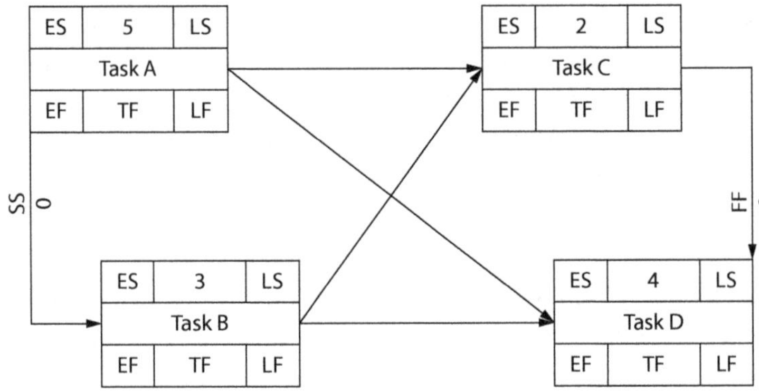

FIGURE 9.6 Precedence Diagram layout.

can be set to 0 days. The Durations of Finish-To-Start sequences, the traditional type of sequence, is always assumed to be zero unless explicitly stated. The initial Precedence Diagram with all needed symbols is shown in Fig. 9.6.

Next, calculate the Precedence Diagram. Set the Early Start of the first Activity, "Task A," to be work day 1. We could envision that this activity begins on a Monday. Given that the durations are listed in "workdays," an Activity with a 5-day duration will start on Monday and end on Friday. Thus, the Early Finish of that Activity would be day 5, not day 6 (which would be the Monday of the next workweek). Also, the start of the following activity would be the day after the completion of the previous activity. The Precedence Diagram's Forward Pass Calculation is shown in Fig. 9.7.

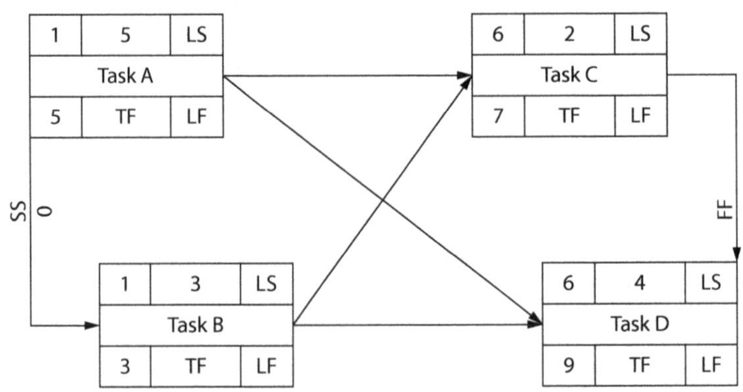

FIGURE 9.7 Forward Pass Precedence Diagram.

Since a finish milestone date was not provided in the problem, information from the Forward Pass is used to set the end of the project. When finishing the Forward Pass, set the Late Finish of the last Activity, "Task D," to be the Early Finish of "Task D." Since the calculation is based on working days, the Backward Pass also considers the impact of calendars. An activity with a Late Finish Date of day 9 and a Duration of 4 will have a Late Start Date of day 6. Once the Backward Pass is complete, finish the CPM calculation by entering the Total Float values. Finally, activities on the Critical Path are highlighted. The completed CPM calculation is shown in Fig. 9.8.

The Fenced-Bar Chart for Prob. 9.1 can be developed by inspection from the Layout Drawing. Both "Task A" and "Task B" are drawn in the upper left-hand side of the chart at the zero point on the x-axis time-scale. One vertical fence can be drawn at the end of the longer of these tasks, "Task A," down the diagram to start the following activities. To the right of this fence are drawn the two following activities "Task C" and "Task D." A fence at the end of the longer of these two activities, "Task D" reveals the completion of the project. The only additional items to include on the Fenced-Bar Chart are the Total Float lines, shown using dashes, between the ends of "Task B" and "Task C" and their following Fences.

Problem 9.2. The solution to Prob. 9.2 should begin as with Prob. 9.1, with a Layout Drawing. The Layout Drawing will have

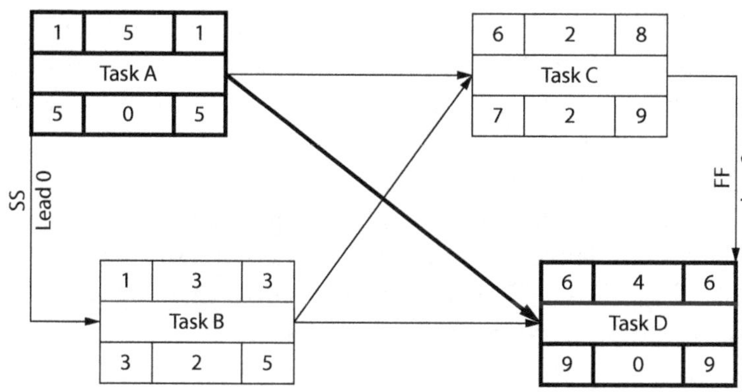

Figure 9.8 Complete Precedence Diagram.

Activity	Prior Activity	Duration (Workday)
A	–	5
B	–	3
C	–	2
D	A, B	4
E	E	2
F	E, F	6

Table 9.2 Activity List for Problem 9.2

the first three activities shown in Table 9.2 on the left and the last three activities on the right. There should be no crossing arrows on the Layout Drawing. By inspecting the Layout Drawing one will be able to see important hints about issues that need to be addressed in each of the types of diagrams he/she will need to produce. For the Precedence Diagram, one can see that there are three starting and three ending activities. This means that one will need to add two Start-to-Start sequence lines at the start of the project, and add two Finish-to-Finish sequence lines at the end of the project. The choice of which to be the starting, or ending activity, is theirs. Some people choose to create a diagram that is symmetric; others select the longest activity as the starting activity. In this matter, there is no one correct answer.

When drawing the Arrow Diagram for Prob. 9.2, there is a hint that Logic Dummies are required. This is because the three activities share some, but not all, of the same set of followers. To draw the most symmetric Arrow Diagram, two Logic Dummies are added to the diagram. These extend from the j-node of "Task B" to the j-nodes of "Task A" and "Task C." By adding the Logic Dummies, the ending events of "Task A" and "Task C" can be kept from affecting "Task E."

Whenever there are Logic Dummies in an Arrow Diagram, the Fenced-Bar Chart solution will require broken Fences. In Prob. 9.2, the fence extending from "Task B" to "Task D" and "Task B" to "Task F" must have a break so that "Task D" and "Task F" do not incorrectly affect "Task E."

Problem 9.3. When completing the Layout Drawing for Prob. 9.3, one will notice that in Table 9.3 "Task B," "Task C," and "Task D"

Activity	Prior Activity	Duration (Workday)
A	–	5
B	A	1
C	A	2
D	A	1
E	B, C	2
F	C, E	1

Table 9.3 Activity List for Problem 9.3

share some, but not all, of the same set of following activities. From this, one can infer that there will be a need for Logic Dummies. The Logic Dummies should be placed between the j-node of "Task C" and the j-node of "Task B." Also one needs a Logic Dummy from the "j-node" of "Task C" and the "j-node" of "Task D." The resulting diagram provides a pair of Logic Dummies that is another common sequence pattern found in many schedules.

Given the need for Logic Dummies, the Fenced-Bar Chart solution for Prob. 9.3 will require broken fences so that, for example, the completion of "Task D" does not affect the start of "Task E."

Additional CPM Diagramming Problems

Problems 9.4 (Table 9.4), **9.5** (Table 9.5), and **9.6** (Table 9.6) are provided without commentary for additional self-study or for use as graded homework problems. For solutions to all problems, please see the footnote on Page 183.

Activity	Prior Activity	Duration (Workdays)
A	–	5
B	A	3
C	B	2
D	A	4
E	B, C	2

Table 9.4 Activity List for Problem 9.4

Activity	Prior Activity	Duration (Workday)
A	–	3
B	A	9
C	A	4
D	C	3
E	C	4
F	B	6
G	F, D	2
H	E, F	3
I	G, H	1

TABLE 9.5 Activity List for Problem 9.5

Activity	Prior Activity	Duration (Workday)
A	D, B	6
B	–	8
C	–	1
D	–	4
E	B, C, F	2
F	–	5

TABLE 9.6 Activity List for Problem 9.6

Resource Scheduling

Problem 9.7. Draw a Fenced-Bar Chart and complete a Limited Resource Allocation of the Activity list shown in Table 9.7. Assume that there are only two resources available per workday. Show all work.

Activity	Prior Activity	Duration (Workday)	Workers Needed per Day
A	–	5	1
B	A	5	2
C	A	3	1
D	B, C	4	1
E	–	8	1
F	B	3	0

TABLE 9.7 Activity List for Problem 9.7

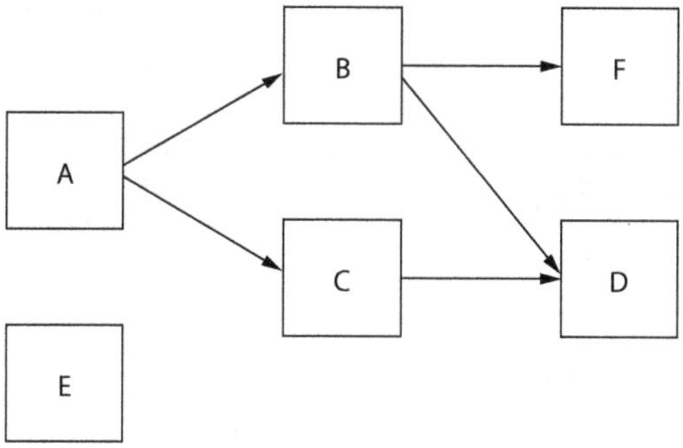

FIGURE 9.9 Layout Drawing.

To begin this problem, develop the Layout Drawing as shown in Fig. 9.9. There are two interesting aspects of the Layout Drawing. First is that "Task B" and "Task C" share some, but not all of the same followers. For the Arrow Diagram, this means that a Logic Dummy may be needed. For the Fenced-Bar Chart, this means that a Broken Fence may be needed. The second aspect of this schedule is that "Task E" has neither a prior activity nor a following activity. In an actual construction schedule, such an activity should be suspected of being incorrect. It is provided for this example, as an example of "what not to do" when defining activities.

Now that the Layout Drawing has been completed, the Fenced-Bar Chart can be developed. As shown in Fig. 9.10, putting "Task F" and "Task D" between "Task B" and "Task C" on the diagram ensures that there are no crossed fences. The broken fence on "Task D" can also be seen on the diagram. To facilitate the resource analysis, the number of resources required per unit time are placed within each activity's bar.

Activity	Duration (workdays)																
Task A	5	1	1	1	1	1											
Task B	5						2	2	2	2	2						
Task F	3											0	0	0			
Task D	4											1	1	1	1		
Task C	3						1	1	1								
Task E	8	1	1	1	1	1	1	1	1								
Calendar (workdays)		1	2	3	4	5	6	7	8	9	10	11	12	13	14	15	16

FIGURE 9.10 Fenced-Bar Chart with resources.

Total Resources	4						1	1	1												
	3						1	1	1												
	2	1	1	1	1	1	1	1	1	1	1										
	1	1	1	1	1	1	1	1	1	1	1	1	1	1	1						
Calendar (workdays)		1	2	3	4	5	6	7	8	9	10	11	12	13	14	15	16	17	18	19	20

Figure 9.11 Early Start Resource Histogram.

With the Fenced-Bar Chart so constructed, the initial Resource Histogram can be made directly by summing the resources needed each day. Figure 9.11 identifies the number of workers needed per day for each day of the schedule before determining the impact of the required limit of two workers per day. On days 6, 7, and 8, there will not be enough workers to complete the scheduled activities.

To conduct the Limited Resource Allocation procedure, begin by establishing the order in which activities will be given access to the limited resource. For this example activities will be sorted in order of increasing Total Float and Early Start. This means that activities on the Critical Path that start the project will be given first access to resources. As a result, resources will be allocated to "Task A" first, followed by "Task B" and then "Task D." Next are "Task F," "Task C," and "Task D." Conducting the Limited Resource Allocation means that the activity on top of the list is added to the schedule if sufficient resources are available, then the activity can start. If insufficient resources, then the activity is delayed until there are sufficient resources. Some delays might push other activities back, but since there are a limited number of resources the delay is justified. Activities that do not require the scarce resource, such as "Task F," are scheduled as early as possible. The result of applying this algorithm, with a limit of two workers per day, and no splitting activities, yields the resource constrained schedule shown in Fig. 9.12.

It is interesting to note that during the initial part of the procedure there is no delay when scheduling "Task A" or "Task B." They are on the original schedule's Critical Path. When scheduling "Task C," "Task D" also gets pushed back because there are insufficient resources available to complete "Task C" until after "Task B" has completed. Completing the schedule is accomplished by pushing "Task E" until later in the

Activity	Duration (workdays)																			
Task A	5	1	1	1	1	1														
Task B	5						2	2	2	2	2									
Task F	3											0	0	0						
Task D	4														1	1	1	1		
Task C	3											1	1	1						
Task E	8											1	1	1	1	1	1	1	1	
Calendar (workdays)		1	2	3	4	5	6	7	8	9	10	11	12	13	14	15	16	17	18	19

Figure 9.12 Limited Resource Allocation.

project. While it is not typically done, one could make the case for adding Activity Logic to the schedule to force the limited resource allocation activity sequence. Doing so would change the Critical Path to reflect that shown in Figure 9.12.

Problem 9.8. In some situations, the limiting factor resources the size of the physical space. Such a case is provided in Prob. 9.8, shown in Table 9.8. This set of activities are in a room that can only accommodate 6 workers per day. What is the shortest amount of time that this project can complete? Show all work.

Problem 9.9. Conduct two Limited Resource analyses using the information found in Table 9.9. First consider only "Resource 1," then consider only "Resource 2." Assume that there are 10 units of "Resource 1" available per workday. Allocate Resource 1 according to a Total Float and Early Start Sort. Assume that there are 14 units of "Resource 2" available per workday. Allocate Resource 2 according to Early Finish and Total Float Sort.

Activity	Prior Activity	Duration (Workday)	Workers Used per Day
A	–	4	2
B	–	2	2
C	A	2	4
D	A	4	6
E	B, D	1	1
F	D	2	2

Table 9.8 Activity List for Problem 9.8

Activity	Prior Activity	Duration (Workday)	Resource 1 Workers Needed per Day	Resource 2 Workers Needed per Day
A	–	4	3	4
B	A, C	9	2	6
C	–	8	6	4
D	A, E	4	2	6
E	–	3	4	2
F	A	5	1	8

Table 9.9 Activity List for Problem 9.9

Problem 9.10. Draw a Fenced-Bar Chart, Arrow Diagram, and Precedence Diagram of a project (shown in Table 9.10) to install three repeated sections of pipe (Activities IDs 2 through 8). Conduct an Unlimited Resource Leveling procedure. Draw a Line–of-Balance diagram of this schedule.

The Unlimited Resource Leveling procedure assumes that there are enough resources to perform the work and that the overall duration of the project cannot be extended. The goal of the Resource Leveling procedure is to reduce the number of times that crews are required to move on and off the site. Resource Leveling is particularly important in repetitive projects such as the one above, where crews of activities not on the Critical Path that are scheduled move on and off the site may not be able to

Activity ID	Activity Name	Prior Activity	Duration (Workday)
1	Mobilize	–	1
2	Survey	1	1
3	Clear and Grub	2	2
4	Excavate	3	1
5	Place Pipe	4	3
6	Weld Pipe	5	2
7	Test	6	1
8	Backfill	7	1
9	Complete Job	8	0

Table 9.10 Activity List for Problem 9.10

return when needed. To answer this problem, begin by creating a Fenced-Bar Chart for the Early Start schedule. Next, assume that there is one resource required for each repetitive activities. To level the resources, move the non-Critical Path activities within their float range to reduce the float among activities of the same type.

The Line-of-Balance schedule is another way to illustrate the importance of Unlimited Resource Leveling. If repetitive activities that have float are scheduled according to their Early Start Dates, as shown in Figure 2.7, the efficient flow of workers through the job will be interrupted. Moving these activities to start at their late starts, similar to the diagram shown in Figure 2.8, will result in one continuously productive line with no breaks.

Time-Cost Scheduling

Problem 9.11. Conduct a time-cost trade-off analysis for the schedule shown in Table 9.11. Show all graphs and charts required as well as the final time-cost trade-off curve.

Problem 9.12. Conduct a time-cost trade-off analysis for the schedule shown in Table 9.12. Show all graphs and charts required as well as the final time-cost trade-off curve.

1. Can activity durations be decreased by changing the crew size, hours worked per day, or equipment used to complete the work? Provide two examples to back up your answer.

Activity	Prior Activity	Normal Time (Workdays)	Normal Cost ($US)	Crash Time (Workdays)	Crash Cost ($US)
A	–	20	1000	10	1300
B	A	20	1500	5	3005
C	–	30	2000	10	2860

TABLE 9.11 Activity List for Problem 9.11

Activity	Prior Activity	Normal Time (Workdays)	Normal Cost ($US)	Crash Time (Workdays)	Crash Cost ($US)
5	–	2	2400	2	2400
10	5	10	5000	10	5000
15	5	15	3000	7	3800
20	10	5	5000	2	6500
25	15	3	1800	3	1800
30	10	9	4500	9	4500
35	20, 25	5	6000	5	6000
40	25	6	3600	4	4200
45	30, 35, 40	4	4800	4	4800

Table 9.12 Activity List for Problem 9.12

2. How, specifically, does decreasing the overall duration of a project increase the overall cost of the project. When might this not hold true? Provide one example for each question.
3. How, specifically, would early completion incentives impact the time-cost tradeoff analysis?

Glossary

Activity The smallest amount of work defined in the Critical Path Method. A work activity is one specific type of work, completed by a single type of crew, in one specific work area. A nonwork activity may also be included in a schedule if that requirement significantly constrains a work activity.

Activity Arrow The basic representation of an activity in Arrow Diagram Method. Requires unique i-node, j-node pair.

Activity Code The set of categories assigned to every activity in a Critical Path Method schedule. Activity Codes are critical for large schedules since scheduling software uses these codes to create customized reports that contain only activities of interest.

Activity Identification The identification number for an activity in the Precedence Diagramming Method. Activity Identifications should contain embedded Activity Codes to clarify and streamline the work of the scheduler.

Actual Cost of Work Performed In construction scheduling, this term refers to the contractor's earnings, not contractor "costs." The cumulative Actual Cost of Work Performed reflects the total contactor earnings since the start of the project. The periodic, or monthly, Actual Cost of Work Performed reflects the contractor earnings during a given progress payment period.

Actual Duration The difference between Actual Finish Date and Actual Start Date. This is used for productivity and claims analysis not for monthly progress updates.

Actual Finish Date The calendar date on which an activity is substantially complete. An activity may be assigned an Actual Finish date even though it has noted deficiencies.

Actual Start Date The calendar date on which work on a given activity began.

Arrow Diagramming Method One of two types of Critical Path Method scheduling techniques. An activity is represented as an arrow. An activity begins at a "i-node" and ends at a "j-node." Sequence between activities is identified by those activities that begin or end at a given node. Also called "activity-on-the-arrow" or simply "arrow diagramming."

Backward Pass The second step in the Critical Path Method scheduling technique. The backward pass answers the question, "How late can an activity start without delaying the project?"

Baseline Schedule A schedule that has been agreed upon by all parties. The originally approved schedule is the first baseline schedule. New baselines will be established during a project following agreements on progress updates and modifications.

Bid Item Code An activity code identifying the owner's cost code associated with the task. Apply one Bid Item Code per activity. Also used during change impact analysis.

Budgeted Cost of Work Performed In construction scheduling, this term refers to the contractor s expected earnings at a given time, resulting from a "Cost-Loaded" "Baseline Schedule."

Calendar The explicit identification of workweek and working time required for Critical Path Method scheduling software.

Changes Variations, additions, or deletions of activities, durations, and sequence that require an update to the baseline schedule.

Claims Changes disputed by the contractor or owner's representative following construction. Mutual resolution of all cost and time impacts of changes during the project is preferred.

Contract Completion Date The date by which the contractor must complete all work. Determined by adding the contract performance period to the Notice To Proceed. A zero duration "Contract Completion Date" activity should be included in schedule with a Finish-No-Later-Than Milestone.

Cost-Loading In construction scheduling, this refers to the proportional assignment of the total construction contract amount to each work activity. Used to identify Budgeted Cost of Work Performed and Actual Cost of Work Performed.

Cost Variance Difference between the Budgeted Cost of Work Performed and Actual Cost of Work Performed. In construction scheduling, this refers to earnings and not cost.

Crashing The action taken when accelerating activities during the Time-Cost Tradeoff procedure. Decreasing an activity's duration typically increases Direct Cost.

Critical Path The set of activities having the minimum total float. This total float is a shared property of all activities on the Critical Path.

Critical Path Method A model reflecting the list of activities, sequence, and duration of tasks needed to complete a construction project. The calculation technique used to determine the Critical Path.

Direct Cost The amount that the contactor must pay for labor, materials, and equipment to complete a specific production activity. Direct costs do not include the cost of project trailers, home office staff, transportation, permits, and other indirect costs.

Early Event Time The earliest that an Arrow Diagram node may complete. A feature of the Forward Pass phase of the Arrow Diagram Method.

Early Finish Date The earliest that an activity may complete. Calculated directly as part of the Forward Pass Phase of the Precedence Diagram Method.

Early Start Date The earliest that an activity may start. A feature of the Forward Pass Phase of the Precedence Diagram Method.

Earned Value The proportion of the total contract amount owed to the contractor for work accomplished. Earnings progress is measured in Percent Complete.

Expended Duration For an in-progress activity, the amount of time crew(s) have worked on that activity to date. Use Remaining Duration, not Expended Duration, for monthly progress updates.

Feature of Work An Activity Code defining a set of tasks with similar subcontractors, crews, or specialized equipment. Required on Corps of Engineers contracts to link Critical Path Method and Quality Control procedures. Also called "Definable Feature of Work".

Finish-No-Later-Than Milestone A date that fixes the Late Finish Date, or Late Event Date. A feature of the Backward Pass Phase of the Critical Path Method. Use to set the Contract Completion Date and other finish milestones.

Finish-On Milestone A date that fixes both the Early and Late Finish Dates, or Early and Late Event Dates. A feature of both the Forward and Backward Pass Phases of the Critical Path Method.

Finish-to-Finish An activity sequence that states that the prior activity must be completed prior to the completion of the following activity. A feature of the Precedence Diagram Method that is typically paired with Start-To-Start sequence.

Finish-to-Start An activity sequence that states that the prior activity must be completed prior to the start of the following activity. A feature of both the Arrow Diagram Method and Precedence Diagram Method.

Forensic Scheduling The systematic process of identifying and classifying activities that deviated from a baseline schedule.

Forward Pass The first phase of the Critical Path Method calculation procedure. The forward pass answers the question "How early can I complete this project?"

i-node The starting node of an activity in Arrow Diagram Method.

Indirect Cost Costs incurred in support of construction work. Including, but are not limited to, the cost of project trailers, computers, telephones, home office staff, transportation, permits, etc.

Initial Schedule The first schedule submitted by the contractor. Once agreed to by all parties, the Initial Schedule becomes the first Baseline Schedule.

Interim Milestone A milestone added between the Notice To Proceed and Contract Completion Date that governs the start and/or end of a Phase of Work.

j-node The ending node of an activity in Arrow Diagram Method.

Lag A duration associated with a Finish-to-Finish Sequence. A feature of the Precedence Diagram Method.

Late Event Time The latest that an Arrow Diagram node may finishe. A feature of the Forward Pass phase of the Arrow Diagram Method.

Late Finish Date The latest that an activity may finish. Calculated directly as part of the Backward Pass Phase of the Precedence Diagram Method.

Late Start Date The latest that an activity may start. A feature of the Backward Pass Phase of the Precedence Diagram Method.

Layout Drawing A simplified Precedence Diagramming Method sketch that helps to reduce crossed Arrow Diagram Method arrows, Precedence Diagram Method sequences, or Fenced-Bar Chart fences when drawing schedules by hand.

Lead A duration associated with a Start-to-Start Sequence. A feature of the Precedence Diagram Method.

Limited Resource Allocation A procedure to determine the earliest that a project may complete, given an insufficient number of required resources.

Liquidated Damages An additional charge, or penalty, incurred for each day that a project completes after the Contract Completion Date.

Logic The sequence between activities or a set of activities.

Logic Dummy Activity A nonwork activity added to correctly represent the sequence among activities that do not all share the same preceding and/or following activities. A feature of the Arrow Diagram Method.

Milestone A date that overrides standard Critical Path Method scheduling calculations. Milestones may fix an activity's Early Start Date, Late Start Date, or both dates. A minimum of two milestones, one for Notice To Proceed and one for the Contract Completion Date, should be included in every schedule.

Node The element that identifies the beginning or ending of an activity. A feature of the Arrow Diagram Method. See i-node, j-node.

Node Numbers The unique number placed within each Node. A feature of the Arrow Diagram Method.

Numbering Dummy Activity A nonwork activity added to ensure that every activity arrow has a unique set of Node Numbers. A feature of the Arrow Diagram Method.

Original Duration The originally estimated time required to complete an activity.

Out-of-Sequence Progress The condition that occurs when an activity begins before its prior activity completes.

Percent Complete The proportion of Earned Value achieved for a given activity, and the overall project, since the start of the project. Should always reported in conjunction with Remaining Duration.

Phase of Work A subset of a project's scope whose start and/or end is defined by contract. Include milestones and hammocks to show Phases of Work.

Plug Date The date used to set a Milestone.

Planning Milestone A zero day duration activity used to track the planned start or finish of a preceding or following set of activities.

Precedence Diagramming Method One of two types of Critical Path Method scheduling techniques. An activity is represented as a node. Sequences between activities are identified by arrows that connect nodes. Also called "activity-on-the-node" or simply "precedence diagramming."

Project An undertaking with a defined start and end.

Project Management The art and science of using available materials, labor, and equipment to complete a project to the satisfaction of all project stakeholders.

Remaining Duration The estimated time required to complete an in-progress activity. Should always be used in conjunction with Percent Complete.

Resource Histogram A graph of resource use with an *x*-axis of time and a *y*-axis of units of total resources required per unit time.

Responsibility Code An activity code identifying the party responsible to complete the task. In addition to contractor and subcontractor codes, owner activities may also need to be identified in the schedule.

Schedule Variance The difference between the Actual Cost of Work Performed and the Budgeted Cost of Work Scheduled.

Sequence The order, or precedence, of activities is identified in a table that lists each activity and the activities that precede (or follow) that activity. Also called "schedule logic" or simply "logic".

Start-to-Start An activity sequence that states the following activity may start after the start of the prior activity. A feature of the Precedence Diagram Method that is typically paired with Start-To-Start sequence.

Start-No-Earlier-Than Milestone A date that fixes the Early Start Date, or Early Event Date. A feature of the Forward Pass Phase of the Critical Path Method. Use to set the Notice to Proceed date and other starting milestones.

Start-On Milestone A date that fixes the Early Start and Late Start Dates, or Early and Late Event Dates.

Time-Cost Tradeoff A procedure used to evaluate the benefit of increasing the cost of some activities to decrease the cost of the overall project. Most effective for projecs with high indirect costs, early completion incentives, and/or liquidated damages.

Time Now The date from which in-progress schedules will be calculcated.

Time Variance The difference between the current schedule calculation date and the date at which the project would have achieved the Scheduled Cost of Work Performed.

Total Float The amount of time an activity may be delayed without delaying the overall project. A property of a series of tasks within a network path.

Unlimited Resource Leveling A procedure to reduce variation in resource requirements.

Work Area Code An Activity Code identifying the physical spaces through which activities must work. When combined with Responsibility Code, may be used to perform simplified Limited Resource Allocation and Resource Leveling.

Work Breakdown Structure The consolidated set of Activity Codes used for one single project

Work Type Code An Activity Code identifying the type of work. Used to identify administrative, procurement, and other non-work tasks. Helpful for identifying and communicating potential owner, tenant, and external delays.

Index

CPSIA information can be obtained
at www.ICGtesting.com
Printed in the USA
JSHW050739040821
17431JS00001B/16

9 780071 849234